Quality on Trial

Bringing Bottom-Line
Accountability to the Quality Effort

Read what these experts have to say about the impact this book and the process it details have had on some of the world's largest and most admired corporations.

"*Quality on Trial* says that if companies listen to their customers and act on what they hear, the payoff is enhanced quality and an improved bottom line. This is a breakthrough book for managers looking for breakthroughs. Read it!"

KENNETH BLANCHARD
best-selling author of *The One Minute Manager*

"I've always said that knowing your customer is the key to success... and this book proves it. An invaluable tool for any company that makes a product or sells a service — and that's every company."

HARVEY MACKAY
best-selling author of *Swim with the Sharks Without Being Eaten Alive*
and *Beware the Naked Man Who Offers You His Shirt*

"As our country enters a new era, becomes more competitive, and increases its appeal as a trading partner, the issue of the quality of our products and services becomes paramount. *Quality on Trial* is a book that should become a cornerstone to every business manager who seeks to strengthen his company's — and his country's — economic viability."

RÓMULO O'FARRILL, JR.
Latin American industrialist and co-founder of the Mexican Businessmen Council

"Having spent a number of years launching and driving our own quality programme, I found this book both highly readable and eminently practical. Where most books on this subject are heavy on theory, this really does give the impression of having been based on real-life experience. My only regret is that it was not available several years ago when I was making a number of mistakes against which it warns."

SIR ANTHONY CLEAVER
Chairman, IBM United Kingdom

Quality
on Trial

Quality on Trial

Bringing Bottom-Line Accountability to the Quality Effort

ROGER J. HOWE
DEE GAEDDERT
MAYNARD A. HOWE

Second Edition

McGraw-Hill, Inc.
New York • San Francisco • Washington, D.C. • Auckland • Bogotá
Caracas • Lisbon • London • Madrid • Mexico City • Milan
Montreal • New Delhi • San Juan • Singapore
Sydney • Tokyo • Toronto

Library of Congress Cataloging-in-Publication Data

Howe, Roger J.
   Quality on trial : bringing bottom-line accountability to the
   quality effort / Roger J. Howe, Dee Gaeddert, Maynard A. Howe.
—2nd ed.
        p.   cm.
   Includes index.
   ISBN 0-07-030583-8
   1. Total quality management. 2. Consumer satisfaction.
3. Organizational change. I. Gaeddert, Dee. II. Howe, Maynard
Alfred. III. Title
HD62. 15.H69 1995
658.4—dc20   94-40245
                                                CIP

The first edition of this book was published by Quality Institute International, copyright 1992.

Quality Institute's Customer Confidence Index™ is a trademark of Quality Institute International. For information contact: Quality Institute International, 444 Cedar Street, 23rd Floor, St. Paul, MN 55101, (612) 220-2404.

1 2 3 4 5 6 7 8 9 0 KP/KP 9 0 9 8 7 6 5 4

ISBN 0-07-030583-8

The sponsoring editor for this book was James H. Bessent, the editorial coordinator was Sarah Dirksen, the text and figure designer was Larkin Mead, the chapter illustration designer was Susan Bartel, the editing supervisor was Jane Palmieri, and the production supervisor was Donald Schmidt. It was set in Garth Graphic by Quality Institute International.

Printed and bound by Kingsport Press.

This book is printed on acid-free paper.

Contents

*This book is dedicated to the staff of
Quality Institute International,
whose energy and commitment have helped
to move us from organized abandonment
to a progressive, successful enterprise.*

Introduction
to the Second Edition

USINESS LEADERS CONTEMPLATING QUALITY INITIATIVES ARE CONFRONTED with a rather uneasy leap of faith. Customers, suppliers, and their own employees expect them to "do quality" and demand that they make the leap, so leap they do, but with little assurance that those quality initiatives will pay off anytime soon. That's the message that first prompted us to write *Quality on Trial. Quality on Trial* was written to answer the business leader's unspoken question: "Is my quality initiative paying off?"

When *Quality on Trial* first appeared, quality for the sake of quality was all the rage. Companies rushed to the quality band-wagon, loudly proclaiming that quality was pursued because it was the right thing to do. But beneath the rhetoric was the unstated expectation that quality and business success are inseparable entities. Improve quality, improve business results. After all, the literature of quality is rife with surveys showing that as companies improve quality, they also improve in critical measures such as market share, sales per employee, return on assets, and the ability to deliver products on time. Few companies questioned the quality mantra and demanded any kind of specific accounting of their quality efforts. Business leaders simply pointed east and justified quality expenditures with "If Japan can, why can't we?"

And in fact, business leaders could point to improved internal efficiencies resulting from their quality efforts and assign cost savings to those efficiencies. They could point to process improvements that resulted in quicker development cycles and faster delivery that they intuitively knew contributed to increased customer satisfaction. Quality appeared to be working, yet appearances proved deceiving. Despite "successful" quality initiatives, companies continued to experience their normal business cycles. And when business was down, the inevitable questions were asked: "Why are we spending all this money on quality? Is our quality initiative paying off?"

Quality on Trial stepped into the void and introduced the Quality Review process — a systematic approach to linking quality initiatives to business results. The Quality Review process provided organizations with a systematic method to collect and use information from its customers for competitive advantage. It provided a means to involve employees in the process of collecting customer information and integrating that information into everyday business activities. It focused quality initiatives on the primary business objective of gaining customer commitment — the lifeblood of any business.

When *Quality on Trial* first appeared, focusing on building customer relationships was a relatively new idea. Companies that embraced the Quality Review process were pioneers. Armed with a basic belief that quality must be integrated into the way a company does business, these companies saw in the Quality Review process a method for achieving that objective. The result has been refocused quality initiatives that have produced improved, measurable, bottom-line business results.

Building customer relationships is the latest business bandwagon. Where once everyone was touting zero defects as the holy grail, today everyone is seeking stronger relationships with customers. But just as companies found as they established quality improvement programs, stronger customer relationships don't just happen. Look at the companies that have been most success-ful at implementing quality improvement programs, and you'll find organizations that took quality theory and systematically integrated it into their way of doing business. Look at quality programs that failed to produce results, and the opposite is true — quality remained a separate function to be performed after the "real work" was done. The same is true for building customer relationships — unless the "business" of building relationships is integrated into the "business" of the business, the effort will not have significant impact on business results.

This second edition of *Quality on Trial* incorporates what we have learned about the Quality Review process as it was implemented in those companies that early on recognized the competitive advantage of building customer relationships. These organizations validated the concept of the Quality Review process, as demonstrated by the business results they achieved. Their successes have been exciting for us, but we recognize that much of it was due to those companies' diligence and creativity in implementing the process. Were there ways, we wondered, to enhance the Quality Review process to make it more responsive to the needs of our customers? The answer was yes.

The concept of the Quality Review process remains un-changed today. What you will find in this edition of *Quality on Trial* is a new tool — TALK^2, an innovative approach to tapping employee expertise to best use information gathered through the Quality Review process. You will also find more examples of how companies have used the Quality Review process to achieve significant business results. As we have said time and time again, business results are what keep the lights on for businesses all over the world.

Quality on Trial is not simply another book on *why* your organi-zation should build better customer relationships. It is a book that describes *how* to build them. It is a book that describes how to turn better customer relationships into business results. The impact of strong customer relationships can be significant, as pioneers of the Quality Review process have proved. The Quality Review process transforms a quality initiative from a leap of faith to a planned, manageable, measurable business initiative.

Foreword:
Organized Abandonment

IN THE FALL OF 1989, WE WERE PRESENTED AN OPPORTUNITY. HAVING recently sold the survey research and quality consulting company with which we had been affiliated for several years, we contemplated starting a new venture. As is the challenge with most new ventures, we sought to build our business on a solid foundation of what we knew, yet not simply clone the business we had just left. Reminded of Peter Drucker's maxim that entrepreneurship is really dependent not so much on a flash of genius as on an "organized abandonment," we took stock of where we had been. When we looked at the work we had been doing for major corporations as a third-party research firm, we knew that we had provided value for our customers. But we also knew that the process — that of gathering information from customers to make businesses successful — could be improved upon.

And so we set about Mr. Drucker's task of organized abandonment. It became clear that we should continue to help organizations gather information from their customers. But we wanted the information to be more actionable. We wanted the process to be more appealing to customers (one that didn't require an incentive to achieve a respectable response rate). We wanted the process to be more appealing to employees, rather than something they felt they needed to defend themselves against. We wanted the process to demonstrate the philosophy we had been preaching to organizations — that it is really their relationship with their customer that is their true competitive edge. We wanted the process to be integral to the business practices of an organization — not a survey project or a quality activity parallel to the "real work" of the enterprise. And, in perhaps our biggest abandonment, we came to realize that we needed to remove ourselves as the objective third party — filtering information from customer to research firm to executive management to employees, at each stage hearing the voice of the customer growing fainter and fainter.

And so our course was charted. We would stay in the quality consulting business, but we would abandon the old, the traditional, in favor of a new, previously untried management system for customer measurement — the Quality Review process. This book has been written to acquaint you with our new approach, the factors motivating organizations to adopt it, and most importantly for you, with the benefits realized when it is implemented.

Our entrepreneurial venture has certainly been gratifying. Developing the process, and the products and organization that support it, has been stimulating. But most rewarding has been the impact on the organizations that have stepped forward to embrace the process. In these organizations a "silent sales force" has emerged to promote the concepts and value of the process for gaining customer commitment — the lifeblood of any business. The process has given new energy to managers, employees and executives exasperated and confused by the frenzied but empty activities many quality initiatives have produced. Corporations, from the world's largest and most admired to the small and relatively unknown, have experienced the benefits of the Quality Review process. It is their positive and enthusiastic response that motivated us to write this book.

But the benefits of this approach, the Quality Review process, are not limited to the private sector. Public institutions, governmental organizations and educational enterprises are also desperately in need of the breath of fresh air that a customer focus can bring. There has been little clamoring from public administrators, politicians and educators to identify their constituents as customers and declare their charter to serve those customers in such a way as to engender their commitment. Our vision extends beyond the private sector, and so *Quality on Trial* has a clear message of change for those organizations that serve our society as well.

The concepts in this book are simple. The process presented here is easy to implement. The impact on organizations of all types is significant. Our sincere hope is that you, too, will benefit from this message.

Quality on Trial

"Business leaders contemplating a quality initiative have been confronted with a rather uneasy leap of faith."

1

Does Quality Pay?
The Jury Is Out

"IT IS BETTER TO FAIL AND UNDERSTAND WHY THAN TO ACHIEVE SUCCESS without understanding." The CEO did not recall the context in which he first heard that remark, but he remembered dismissing it as one of those slick-sounding but irrelevant statements made by people who didn't understand the reality of the business world. Success was the yardstick; that was all the understanding required — at least that was the CEO's thinking before this day's board meeting.

The CEO picked up the hard copy of the "State of Quality" presentation he had made to his board of directors. It was a good-news presentation. For the third straight year, the results of the company's annual customer survey showed an across-the-board increase in customer satisfaction. Communication efforts had increased quality awareness. Quality improvement was on the upswing. Statistical process control was implemented in nearly all of the company's worldwide manufacturing facilities. Hours of quality training were up 50 percent from last year, and quality improvement team activity was up 25 percent. Nearly a dozen quality function deployment projects were underway. All of the company's manufacturing facilities were working toward ISO 9002 certification, and many facilities were self-assessing against the Malcolm Baldrige Award criteria.

The quality initiative appeared to be a success. The CEO's presentation was a big hit. The board was enthusiastic. They understood that quality was important to the future of the company. They agreed: No compromise of quality could be tolerated if the company were to maintain its competitive advantage. But despite all the positive reaction to his presentation, the CEO was nagged by one question: How were these quality activities of which the company was so proud linked to business results? Specifically, could the company's quality programs be linked to the revenue and market share objectives the CEO was driving through the sales and marketing organizations?

The CEO was convinced his quality initiative was successful, but when asked to justify the quality investment in terms of business results, he couldn't. In fact, as he reflected on the question, the CEO realized he rarely brought up the quality program when talking to his sales and marketing people, save some general references to customer satisfaction.

"What *was* the link between quality and business results?" wondered the CEO. Internal process improvement was saving the company money — that was obvious — but shouldn't there be a return on the quality investment that went beyond just reducing defects? Shouldn't quality be linked to the company's bottom-line business results? Shouldn't quality contribute to the growth of business?

For the first time in his corporate career, the CEO realized the consequences of a success not fully understood. Intuitively, he knew the quality initiative was on the right track. He labeled it a success, but he didn't understand why. He couldn't link the quality initiatives, the mainline activities of his organization, and the company's objectives of increased revenue and market share. Further, he realized that without a system for doing so, there was no way to make those connections with any certainty.

The CEO tossed the copy of his presentation in his briefcase. He made a note for the agenda of his next staff meeting: "For discussion: Is our quality initiative paying off?"

Denominator Management

For the CEO just described, the jury is still out on the question "Does quality pay?" And he is not alone. Faced with less-than-spectacular profits, many companies are realizing that internally focused activities such as cost reduction and the ever-popular "rightsizing" are necessary to remain competitive, but they are not sufficient. Reality is setting in: Only so much profit can be gained from squeezing the waste and inefficiency from an organization. In the end, a company must grow its revenue to survive.

Much has been written about attempts to change the priorities of the modern corporation. However, as the hundreds of line executives we have interviewed are quick to point out, none of these changes, including the emphasis on quality, is possible if the corporation is not profitable and doesn't provide a reasonable return to its shareholders. When all is said and done, profit is what keeps the lights on.

General consensus in the business world is that profit is driven by increased productivity. But as Gary Hamel of the London Business School notes, productivity is a ratio — net income over some investment. Thus, there are two ways to effect changes in productivity and two ways to increase profits. Profits rise when the investment required to generate income goes down or when revenue rises. Invariably, acute pressure to increase profits puts emphasis on the denominator of the productivity ratio. Reducing investment (which is achieved through improved operations) while maintaining the current level of income yields increased productivity and profits.

"We have produced a generation of managers," says Hamel, "who are the world's best at denominator management,"[1] and who saw in the new emphasis on quality yet another way to manage costs. Quality programs have, after all, generally been applied in an effort to reduce the denominator of the ratio. Research on the quality programs of America's top 1,000 corporations suggests that

"As a result of employees conducting Quality Reviews with our largest customers, our quality improvement initiatives are now being driven by our external customers. We no longer waste time on extraneous issues that provided no value to our customers and small savings, if any, to us. We can point to our bottom line and identify dollars that we have saved as a result of doing things our customers want — not things we thought were 'nice to do.'"

Dan Dougherty,
General Manager,
Cargill North American
Flour Milling

"most companies were forced to join the quality revolution [not for altruistic reasons, but] due to a crisis within their industry or company."[2]

Numerator Management

Is denominator management the right long-term management strategy for corporate success today? Denominator management produces profitability not through working with suppliers to lower their costs, not through partnering with customers to create new products and new businesses, and not through tapping the minds of employees for new ideas. In fact, Hamel argues, in order to increase productivity and profitability, companies have done just the opposite — squeezed suppliers, reduced service to customers, and downsized the work force.[3] Increasing the numerator of the productivity ratio is the more visionary approach — in other words, leveraging existing resources to increase profits by increasing revenue.

Numerator management represents the kind of offensive business strategy needed in today's highly competitive, global business environment. As necessary as restructuring, downsizing, cost-cutting, and reorganizing may be, none of these tactics is sufficient for long-term business success. Long-term success (and some may argue survival) depends on the ability to grow — not just managing operational expenses (denominator management), but growing the company's revenue (numerator management).

"Each quality demigod ... boasts his own set of commandments, rituals and disciples. And within each approach, corporate managers are confronted by a numbing maze of acronyms and buzzwords."

Business Week
1991 Quality issue

Businesses should always looks for ways to cut costs, eliminate redundant operations, and flatten bureaucracy. However, a growth strategy demands that the company's primary focus be external — on customers, suppliers, and the industry in general. Internally, a company should not focus exclusively on processes; it must tap and capitalize on the ideas of those who make the processes run — the organization's people.

Most companies intuitively recognize the importance of the numerator in the profitability equation, and they set specific objectives for revenue growth. However, close examination of their strategies often shows that the growth "plan" is totally dependent on becoming more structurally competitive. These companies seem to assume that if the company structure is "correct," growth will inevitably follow. In other words, they manage the denominator, not the numerator. It is in this context that the generic "quality improvement" banner is unfurled.

Mixed Evidence

With evangelical fervor, quality gurus and their disciples have carried the quality message and a plethora of quality techniques into thousands of major corporations. Supported by such slogans as "Commitment to Quality" and "Zero Defects" and quality activities ranging from self-managed work teams to i-dotting, t-crossing, detailed work processes, quality programs are proliferating faster than fat-free foods. Spurred by the need to be "customer focused," "market driven," and "obsessed with the customer," business leaders are turning to third-party research firms with the plea, "Tell me if my customers are happy." Quality tools and techniques abound: the Seven Old Tools, the Seven New Tools, SPC, QFD, TQM,[4] and just about any other combination of three letters you might imagine.

The Voice of the Customer

"Customer-driven strategy is truly the key to the Japanese 'miracle.' The message is actually deceptively simple: *Bring the customer's voice in-house* so you can build your policies and your products around your customer's needs. Japan's leading manufacturing and service companies use this strategy, while American executives struggle to copy control charts and fishbone diagrams to master the quality-driven strategy that Japan has long since outgrown.

"Customer-driven strategy requires quality *and* service excellence at every level. In the process, customers are brought into every department of the company, and their voices heard and acted on. The long-term result of this approach is keeping customers for life."

J. Koob Cannie
Keeping Customers for Life, 1991

Clearly, an abundance of tools exist to implement quality improvement — but what does "improving quality" really mean? Is quality improvement merely another way of managing the denominator, of doing more of the same with less? Or does quality play a role in managing the productivity numerator, in growing the revenue? Ultimately, the question must be addressed: Are quality initiatives paying off?

It is, we believe, a sad commentary that most business leaders cannot positively address the linkage between quality and business results. The good news is that they are looking more critically at their quality initiatives, and they are demanding an accounting.

On a macro level, it appears that quality pays. In *The PIMS Principles,* authors Buzzell and Gale note a positive relationship between customers' perception of a company's quality relative to its competition and the company's return on investment. Quality, say the authors, is "the most important single factor affecting a business unit's performance." In the short run, superior quality yields increased profits via premium pricing. PIMS businesses that ranked in the top third on relative quality sold their products or services, on average, at prices 5 to 6 percent higher (relative to the competition) than those in the bottom third.[5]

Further evidence that quality pays: A survey by the General Accounting Office found that the top 20 scorers in the competition for the Malcolm Baldrige National Quality Award have demonstrated some improvement in critical measures such as market share, sales per employee, return on assets, and the ability to deliver products on time.[6]

But appearances can be deceiving. As the soul-searching CEO came to realize, making the link between a successful quality initiative and business success is not always easy. He is not alone.

"Quality in most organizations is an activity parallel to the way a company conducts its daily business. Most quality initiatives have not been institutionalized — have not been integrated."

Paul Noakes
Vice President of Quality
Motorola

A 1989 Gallup survey performed for the American Society for Quality Control (ASQC) found that only 28 percent of executives interviewed say they have achieved significant results from their quality initiatives — "significant" defined as increased profitability or market share. That is down from 46 percent in 1987.[7] A similar study by the British Department of Trade and Industry revealed that in 86 percent of companies examined, the design and implementation of advanced manufacturing technologies and systems failed to realize any improvements in flexibility, quality, throughput time or due-date performance. Amazingly, *43 percent of these companies did not improve their overall competitive position.*[8]

Total Quality Management (TQM) programs, which emphasize reducing defects in internal processes and identifying and satisfying customer requirements, are the most popular of the current quality approaches but are not as successful as they should be, according to a survey conducted for the American Electronics Association. Of 311 companies studied, 63 percent have achieved less than a 10 percent reduction in defects from their in-house programs, and 72 percent achieved less than a 10 percent reduction in defects from their supplier programs.[9]

Even scoring high against the Malcolm Baldrige criteria, or actually winning a Baldrige Award, is no guarantee of success. Baldrige Award winners have seen their stock prices fluctuate, experienced product development delays, and lost money on new ventures.[10] Small-business winner Wallace Company (1990) laid off more than a quarter of its work force and flirted with bankruptcy.[11]

The Leap of Faith

Faced with mixed evidence of a link between quality initiatives and positive business results, overwhelmed by the plethora of quality philosophies, tools and techniques, and pressured by the "political correctness" of visibly supporting anything and everything that has to do with quality, business leaders contemplating a quality initiative have been confronted with a rather uneasy leap of faith. Customers, suppliers, and their own employees expect them to "do quality," and demand that they make the leap, so leap they do, but with little assurance that the quality initiative will pay off anytime soon. The common rationale for many quality initiatives is that they will pay off "five or six years down the line," and the business executive can only hope that shareholders are willing to wait. "I'm not very patient," noted one CEO. "You talk about five years, seven years [to get a quality payback], and I get very frustrated with the pace."[12]

Do companies today have a quality problem? We believe the answer is an unequivocal "yes," but not for the reasons commonly cited. The quality problem is less an indictment of the quality of products or services than one might expect. Nor is the problem entirely, as some have suggested, a crisis in the quality of business leadership. Quite simply, the quality problem is that too few business executives, middle managers and employees know if all of the money, all of the time, all of the resources, and all of the frantic activity their organizations are devoting to quality are having a favorable impact on business results and customer commitment.

That statement is not intended to accuse business leaders of either incompetence or negligence. Quite the contrary. Whether for altruistic reasons or hardball business advantage, most business

"Until a firm connection between continuous improvement and the bottom line is made, measured and regularly reported to senior management, the people in the executive suite are not going to take quality seriously."

J. Bowles and J. Hammond
Beyond Quality, 1991

"It is true that almost all quality-oriented organizations do customer surveys. But too often the survey is conducted, results are tabulated, and then nobody knows what to do. There is no structure that ties customer satisfaction to internal business processes. The customer side of quality is seldom linked effectively to the process quality improvement effort."

California Management Review
Spring 1993, Vol. 35 – No. 3

leaders want to improve quality and satisfy customers. What is missing is the cause-and-effect link between quality activities and ongoing business plans and practices. In fact, many quality initiatives are parallel activities that never become fully integrated into the everyday business of the company.

For example, take the key criterion used to validate the success of one company's quality initiative — total number of documented work processes. By this measure, the company's quality initiative was certainly successful. A steadily increasing number of processes was documented — 1,500 at the peak. However, a closer look found that several departments were documenting processes with little or no impact on the key business measurements of the company. In fact, several major processes with the potential to impact the business went undocumented and unimproved. Why? People were rewarded for documenting processes — any processes, not necessarily those tied to specific business objectives. Consequently, processes that were easy to document were tackled first because more could be "accomplished" with less time away from "real work."

Another company based the "success" of its quality initiative on the number of teams formed, number of quality meetings attended, and the number of quality suggestions received. Although these measures may have some value, they give no indication of whether a company's quality initiative is linked to fundamental business results. It is the lack of a strong link between quality initiatives and business results that differentiates the quality initiatives that fail from those that succeed. In other words, if your quality initiative is to pay off, your quality processes must be *integrated* with ongoing business practices. But that somewhat ideological statement begs a second question — "How do I integrate my quality initiative into ongoing business practices?" The balance of this book is dedicated to providing the answer.

*Q*uality on Trial is for every business executive who has ever, in the midnight of the soul, had doubts about the contribution of his or her company's quality initiative to business results. It is for all middle managers who have received memos from the powers that be inciting them to "do quality" while maintaining production, inventory, revenue, and cost objectives. It is for every frustrated employee who has ever been told that the new company policy is "do it right the first time," as if he or she hadn't been trying to do things right all along. It is for every customer who has ever filled out a multi-page customer satisfaction survey from a supplier only to find that nothing about the supplier's performance ever changes. It is for everyone who has ever wondered, "Is all this quality stuff paying off?"

In this book, we present an approach to linking your quality initiative to business results. We present a numerator approach to managing quality that can be directly linked to increased revenue. We provide you with a customer management system that is innovative, has proved successful, and is easy to implement. We call this approach the "Quality Review process." It is designed to help you integrate your quality initiatives into everyday business practices and to establish a cause-and-effect relationship between quality and immediate and long-term business objectives.

That is not to say the Quality Review process is a one-tool-solves-all approach to quality. Earlier we noted that most companies are forced to join the quality revolution due to a crisis within their industry or company. This "crisis quality" mentality is significant in that it pushes an organization in many directions at the same time, in ways unique to an environment in crisis. Consequently, all corporations do not operate under one consistent quality theory; their quality programs mix philosophies and use a variety of tools.[13]

Which approach is right for *your* quality initiative? Many tools, techniques and philosophies are valuable, maybe indispensable, to producing quality products and delivering quality services. This book is not intended to replace those tools. What the Quality Review process will do is take the uncertainty out of your leap (or further leap) into quality by providing a systematic, customer-focused approach for determining which tools are right for your company in the context of your business objectives.

Obviously, your ultimate objective is an affirmative answer to the question, "Is my quality initiative paying off?" Equally obvious: if the answer is to be "yes," then there must be a tight link between the management approach, the quality process, and business results.

Chapters 2 through 5 define the key elements that form the link between quality and the everyday activities of your business. What organizational focus is reflected in your mission and vision, and how do you communicate that purpose to your employees, suppliers and customers? What is your vantage point and perspective on your customers and your employees? Are you stuck at the top of a Ferris wheel? What is the state of your relationships with your customers and employees?

Chapter 6 provides a detailed, step-by-step approach for establishing a customer management system (for both external and internal customers) that helps link quality to business processes and that provides the customer information required for a sound and effective relationship management strategy. In Chapter 7, the same principles are applied to establish a relationship management system between managers and employees. Chapter 8 reapplies the principles to supplier management.

While Chapters 2 through 8 provide you with the raw materials for integrating quality into your business, Chapters 9 and 10 provide you with a picture of integration: what kind of business results should be expected and demanded from a quality initiative (Chapter 9), and what steps the business leader must take to ensure quality is linked to business results (Chapter 10).

If the Quality Review process can provide the missing link between quality and business results, why isn't every company using it? In Chapter 11 we'll meet Hazel and find the answer. In that answer you'll find an opportunity for competitive advantage.

1   Gary Hamel. "Corporate Imagination and Expeditionary Marketing." Presentation to The Masters Forum, September 9, 1993. The Masters Forum, 5620 Smetana Drive, Suite 270, Minnetonka, MN 55343.

2   Phillip E. Kendall. *Vision 2000: America's Top 1,000 Companies' Quality Progress.* GOAL/QPC Research Committee 1990 Research Report No. 90-04-01. Copyright 1990 GOAL/QPC, Methuen, MA.

3   Hamel. *Ibid.*

4   SPC (Statistical Process Control), QFD (Quality Function Deployment), TQM (Total Quality Management). The Seven Old Tools are cause-and-effect diagrams, pareto charts, histograms, check sheets, control charts, bar graphs, and scatter diagrams. The Seven New Tools are relations diagrams, affinity charts, systematic diagrams, matrix diagrams, matrix data analysis, process decision program chart (PDCA) and arrow diagram.

5   Robert D. Buzzell and Bradley T. Gale. *The PIMS Principles.* Copyright 1987. The Free Press, a division of Macmillan, Inc., New York, NY. The PIMS (Profit Impact of Marketing Strategy) Program was initiated in 1972. Since that time, more than 450 companies have contributed information documenting the strategies and financial results of nearly 3,000 strategic business units for periods that range from 2 to 12 years.

6   *Management Practices: U.S. Companies Improve Performance through Quality Efforts.* United States General Accounting Office, GAO/NSIAD-91-190. May, 1991.

7   *Quality: Executive Priority or Afterthought?* A survey conducted by The Gallup Organization for the American Society for Quality Control. Copyright 1989 American Society for Quality Control, Milwaukee, WI.

8   *Manufacturing, Organisation, People and Systems.* British Department of Trade and Industry, 1991.

9   Data copyright 1991 Pittiglio, Rabin, Toup and McGrath and KPMG Peat Marwick High Technology Practice.

10   "Is the Baldrige Overblown," *Fortune,* July 1, 1991.

11   "The Ecstasy and the Agony," *Business Week,* October 21, 1991.

12   Where Did They Go Wrong," *Business Week,* October 25, 1991.

13   Phillip E. Kendall. *Vision 2000: America's Top 1,000 Companies' Quality Progress.* GOAL/QPC Research Committee 1990 Research Report No. 90-04-01. Copyright 1990 GOAL/QPC, Methuen, MA.

"Leaders who are not sensitive to the social movement aspects of quality are led by the movement — they do not lead it."

2

Why Quality Initiatives Fail: The Evidence

A DRAMATIC EXAMPLE OF THE IMPACT OF A LACK OF INTEGRATION between quality activities and business objectives occurred within a large exporter of computer chips to Japan. The company had achieved world-class quality leadership for its product and was convinced that if it developed a chip with a 15-year guarantee, it would dominate the market through the '90s. With an internal focus on work processes, new manufacturing technology, statistical process control training for all employees and a gain-sharing plan based on results, this company turned itself inside out for two years, spent $150 million and yes, created the ultimate, 15-year, defect-free computer chip.

Unfortunately, the company's customers didn't want a chip with a 15-year life. They wanted a chip with a 2- to 3-year life that could be removed from a circuit board and replaced with the next generation chip without replacing a major part of the hardware. The chip manufacturer and its customers were out of sync.

Did the chip manufacturer improve quality? Definitely. Did the company have the "best" product? In technical terms, without question. But did the product meet customer needs? No. Did the quality effort lead to customer loyalty? No. Did the company get the business? No — the business went to competitors who were out there listening to the voice of the customer.

The problems of the chip manufacturer are illustrative of a key conclusion of the research we have conducted over the past 20 years working with Global 1000 corporations, conducting surveys, interviews and focus groups with employees, customers, suppliers, executives, line managers and quality professionals in major industry groups. We have concluded that most companies' quality initiatives focus primarily on internal processes that are rarely linked to desired business results.

This chapter will review the considerable evidence that supports that conclusion, but it will not merely lament that finding. A second conclusion of our research, that quality has taken on the characteristics of a social movement, *provides a framework for understanding what it means to integrate quality into the everyday activities of the business.* Based on that understanding, it is possible for an organization to implement a management system that links its quality initiative with business results.

Focusing Inside Out

The chip manufacturer represents the philosophy that a quality initiative's proper focus is on improvement of key internal processes. "If only we improve our processes, then we will have higher quality," reasoned its management. And so management identified "key" business processes (from an internal perspective) and set about defining measurements, establishing objectives, and implementing programs to achieve those objectives — all activities applauded by modern quality gurus.

However, we contend that when the *primary* focus of quality measurement and reporting is on measuring participation in activities (number of hours of training, number of team meetings, number of suggestions, and the like) rather than on the impact of those activities on business results, the company's time and money is wasted. A color graphic displaying the number of hours spent in quality meetings and training sends the confusing message to employees that quality is an activity separate from the real work of the organization.

The chip manufacturer is not alone in its focus on internal quality activities. The top methods for improving quality, as executives view them, are motivation, leadership and education (Figure 2.1) — activities that are internally focused. Notable by its absence is "customer relations." We say notable because the most important factor in continuing to do business with a supplier, as *customers* view it, is the quality of the relationship between the customer and the primary contact in the supplier's organization (Figure 2.2).

Further support of this finding: 68 percent of customers who leave a supplier do so because of an attitude of indifference from one or more individuals in the supplier's organization.[1] They leave not because of price, not because of product quality or delivery, but because they perceive that the supplier is indifferent to their needs.

When executives have an inside-out focus, it is not surprising that employees have the same view. Our research shows that only a small number of employees know who their customers are. Another startling finding: Of employees surveyed, only 17 percent feel they understand their customers' requirements; only 52 percent feel communication with

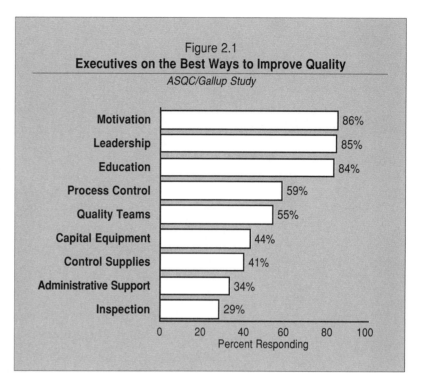

Figure 2.1
Executives on the Best Ways to Improve Quality
ASQC/Gallup Study

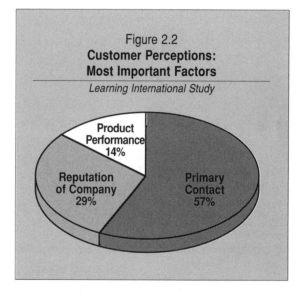

Figure 2.2
**Customer Perceptions:
Most Important Factors**
Learning International Study

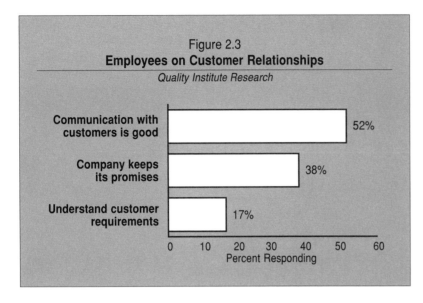

Figure 2.3
Employees on Customer Relationships
Quality Institute Research

Communication with customers is good — 52%

Company keeps its promises — 38%

Understand customer requirements — 17%

Percent Responding

customers is good; and only 38 percent feel their company keeps its promises to customers (Figure 2.3).

Companys' focus on internal processes is not surprising. Harvard professor Theodore Levitt notes that inside the organization is where the work activity occurs, where budgets and plans are developed, where performance is evaluated and where efficiency is continually emphasized. It is inside the organization where one receives rewards, develops friendships and alliances, and where changes in policies, procedures and organization structure are in the control of management and employees. This internal focus is based on the belief that what takes place inside the organization is more important than conditions outside of the organization.[2]

Sometimes, this internal-focus philosophy gets carried to ridiculous extremes. Consider the case from Great Britain of a local transit authority who, when told that bus drivers were "speeding past queues of people with a smile and a wave of the hand," replied, "It's impossible for drivers to keep their timetables if they have to stop for passengers."[3]

Although it may sound silly, the story of the transit authority is not all that different from that of the chip manufacturer, who spent $150 million to produce something customers didn't want. Both the transit authority and the chip manufacturer made the same common mistake of formulating their quality initiatives in terms of internal processes and procedures. Regardless of their official stand on the importance of meeting customer needs, they left the customer out of the quality initiative, figuratively and literally.

Granted, the quality movement has spurred some attempt to change organizations from an inside to an outside focus. However, despite the hoopla surrounding "customer focus" and "market-

"Quality has never been integrated into every process in the business. Executives are launching company-wide efforts to encourage all employees to improve themselves and their job process, but what is needed is a formal system throughout the company for integrating quality into business process."

A. Gunneson
Quality, 1987

driven quality," quality in many corporations is still primarily an activity-oriented, internally driven effort. Companies communicate the need to change, but they lack a system to energize *behavior* change in the organization.

What Are the Consequences?

Qualitative information gathered through our interviews with business leaders as well as analysis of quantitative research makes a strong case that quality initiatives are not delivering on their potential payoff. The chip manufacturer and the transit authority are not alone in this failure. Time and again we have seen companies diligently stow all of the corporation's quality eggs in a single guru's basket, or invest exclusively in one or two quality approaches, which are then force-fed across the organization.

Again, it is not our intent to discredit any one single approach to quality. Rather, our point is that regardless of which guru or what tools are employed, companies that put all their quality eggs in one basket have the same worldview of quality; that is, that quality is the improvement of internal processes and procedures in order to meet customer requirements. Their primary focus is not on the requirements, but on the processes. Their focus is not external but internal. It is the *context* of their quality initiatives, not the *content*, with which we take issue. It is the context of a quality initiative that is addressed by the Quality Review process.

How does one recognize an internally focused quality initiative? Not by the activities going on, but by the way those activities are positioned vis-à-vis the ongoing business plans and practices of the corporation.

The internally focused organization will generally proclaim its conversion to quality in the voice of the zealot. There is a lot of talk about "visions" and "missions" but very little concrete direction. The initial focus of the quality initiative is on slogans and banners intended to increase quality awareness, leaving to the individual employee the alchemy of transforming awareness into action. Does this approach create awareness? People *are* more aware of quality issues, but in terms of positive impact on the

"Despite general business claims that customer satisfaction is of paramount importance, findings indicate that only 22% of U.S. companies use customer measures in their planning process."

American Quality Foundation and Ernst & Young International Quality Study, 1991

*"Quality initiatives
fail because they are
delegated by top
management to staff
groups, the human
resource function or
a quality function.
These staff groups
design the program
and own it. Line
management does
not. Also, staff groups
are not always held
in high esteem. They
are not thought to
know the business of
each unit well
enough to be
helpful."*

M. Beer
Quality Progress
February, 1988

organization, the "all hat and no cattle" approach to quality only reinforces the notion that quality is a superficial activity parallel to meeting daily business requirements. We have found that more often than not, empty slogans trivialize the quality initiative and, in fact, do more to alienate employees than they do to motivate employees (Chapters 3, 4).

Internally focused organizations steal shamelessly. Used as a tongue-in-cheek synonym for "benchmarking," "stealing shame-lessly" has come to mean the process of identifying best practices and adapting them to one's own organization. Unfortunately, some-thing is lost in the translation. Benchmarking certainly can be a valuable quality activity; however, too many companies try to use benchmarking as a quick fix. They steal a process here, steal a methodology there, understanding neither their own processes nor the process they are attempting to overlay on their business prac-tices. In that context, "stealing shamelessly" (to say nothing of the ethical implications of the phrase) damages an organization more than it produces benefit.

A direct consequence of failure to integrate quality into ongoing business practices is that as quality activities increase, so does the quality staff. Administration of all the quality activities requires companies to build large staffs of quality professionals. In turn, the proliferation of quality professionals communicates a message that quality is the responsibility of a designated department rather than of everyone in the company. Operations people note a curious correla-tion between the number of quality personnel, the meetings and seminars employees are asked to attend, and the amount of "vital" paperwork to which they must devote their time instead of looking after the requirements of the business. One company's employee quality assessment noted, "It seems as though all we really have is more paperwork and many contrived, meaningless, quantitative measurements that supposedly ensure quality. Quality is much, much more than quantitative things that can be measured."

A corollary to the large staff approach to quality is augmenting the internal program with outside consultants. Many companies hire outside consultants as the *primary* drivers of their quality initiatives, only to find that the consultant's impact is short-lived. In order to maintain momentum, the consultant must be a continual presence — the quality effort is never internalized by employees or management. In a consultant-driven quality initiative, employees and managers develop a passive resistance to the consultant and his/her program — placating them while taking the silent position, "And this, too, shall pass."

Team formation is another common basket for the internally focused corporation's quality eggs, regardless of whether the team approach is applicable to the situation at hand. Allow us a favorite story here: A manager once told the personnel office that he needed to hire a highly intelligent person for a project — "someone with an IQ of 150." Personnel couldn't find such a person to hire, so they put together a team of three people, each with an IQ of 50. The moral of the story: teams are not the solution to every problem. Nor is consensus (a corollary of the team syndrome) always a requirement. Teams can prove valuable, but only if they are used in situations where two heads are *truly* better than one. When a single individual has both the knowledge and authority to make a decision, it is a serious waste of resources to form a team and strive for consensus. Also, teams cannot function in a vacuum; they must have viable information for making decisions, which implies that there is a system in place for gathering required information (Chapters 6, 7, 8).

Universal quality training is another commonly perceived cure-all for internally focused companies' quality initiatives. Many quality professionals have inadvertently led their organizations astray by promoting the need to train everyone before "real" quality improvement can take place. Attempts to change the company's culture by changing attitudes through education (rather than by altering behaviors by providing a system for behavior change) do more to delay the achievement of business results than to realize them. Only recently are some organizations becoming seriously concerned about the cry they are hearing from their people — "We have to put aside our work, it's time to go to a quality class." This, indeed, is a telling commentary on the wasted potential of training as a valuable resource to a quality initiative.

"Companies now shell out $750 million a year to 1,500 third-party providers of advice and materials on quality.... No wonder, then, that nearly every consultant has added quality to its mix of advice."

Business Week,
1991 Quality issue

Certainly, training can be effective in helping employees shift their focus from inside to outside, but our research finds that few corporate training curriculums are designed to accomplish this objective. We have found that customer relationship management (outside focus) is either nonexistent or is taught as theory and not as a system that the organization has in place to move it from an inside to an outside focus.

Management believes quality education is important (Figure 2.1, page 15), and research confirms that management is backing this belief with actions: 82 percent of the companies in that American Society for Quality Control study had specific quality education programs in place.[4] However, in the majority of companies we have studied, this rapidly proliferating "quality training" is nothing more than existing training, such as team building or problem solving, repackaged with a quality title. The training (which in the proper context has a great deal of value) is made less efficient when presented to managers and employees with a complicated, confusing array of quality jargon — that is, when it is presented as something outside normal business practices. And what is employee reaction to this education? Not even one in five say the quality training they receive is relevant to their jobs (Figure 2.4).

All of the above activities certainly have a place in quality initiatives. All have been implemented successfully, and many companies can directly relate their activities in these areas to productivity gains, cost reductions and other business results. However, we submit that those successes only scratch the surface of the potential quality payoff available to those organizations that switch the context of those activities from internal to external. When one examines recognized successes, one generally finds that they are isolated in nooks and

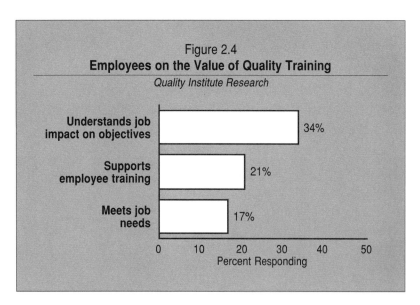

Figure 2.4
Employees on the Value of Quality Training
Quality Institute Research

Understands job impact on objectives — 34%
Supports employee training — 21%
Meets job needs — 17%

Percent Responding

crannies of the organization —
that when the quality initiative
is viewed as a whole, it appears disjointed, segmented and
inconsistent. What better
witness for this assertion than
employees themselves?

When the question is quality, employees' commitment
and their perceptions of management credibility are less
than stellar. Our research finds
only 43 percent of employees
believe that management is

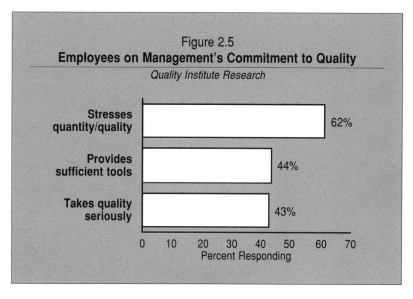

really serious about improving quality (Figure 2.5). We are not
alone in observing that there is a very evident lack of employee
confidence in management's quality direction. A Gallup Organization survey performed for the American Society for Quality
Control concluded that there is a serious lack of credibility in most
companies' quality initiatives. Quality programs have not en-
gaged or are not available to everyone; some people are turned off
by what they perceive as a gap between the company's talk and
action on quality; and they are dissatisfied with the rate of quality
improvement achieved to date.[5]

This is an unfortunate perception because, we believe, it is
simply not true in the majority of cases. Management *is* committed
to quality, and management *is* doing the right things, but the
message is not getting through. We feel the reason is that most
quality initiatives are internally focused, which tends to fragment
the effort as departments and work groups optimize their operations at the expense of the larger enterprise view. This fragmen-
tation, and subsequent lack of impact on the customer, is viewed
by employees, customers and the general public as caused by a lack
of management commitment.

We believe most organizations have not recognized the internal/
external context distinction or its consequences. Of those that have,
many do not have a system for developing the external perspective

that would unify their quality initiatives. Their approach to quality remains reactive and theoretical, as opposed to visionary and behavior-oriented.

Integrating Quality

We may lament or rationalize findings such as those just described, but we are better served by looking for the root cause of the problem. As stated earlier, part of the problem lies in the lack of a management process for integrating quality into daily business practices. But before one can adequately address the management process issue, one ought to have a clear understanding of what integration of quality with ongoing business practices really means, which leads to a second major conclusion of our research: We believe that quality has taken on the characteristics of a social movement.

Like the Labor Movement, the Civil Rights Movement or the Women's Movement, quality has taken on a social significance that goes beyond implementation of quality processes and procedures in the corporation. Primarily because of the astounding success of Japan's quality-based economic resurgence, consumers around the world have redefined their expectations of the products and services for which they are willing to pay. The quality movement has spurred people to redefine their personal values and attitudes toward quality and has significantly altered their behavior.

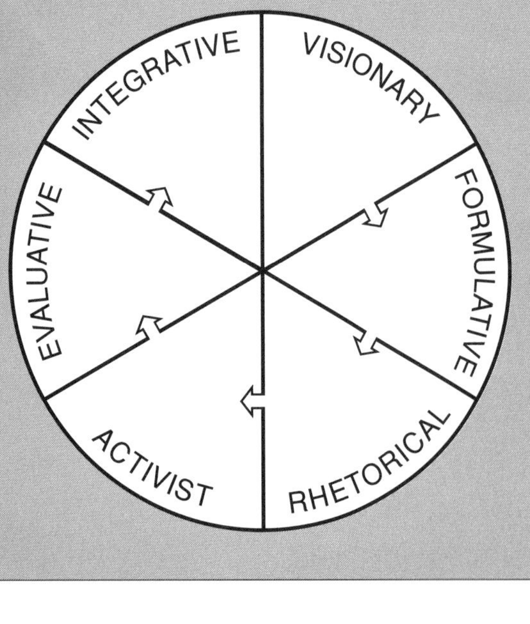

Studying the phases of social movements can help us understand the quality movement. The ultimate goal of a successful quality initiative is the same as the goal of a social movement — to have its concepts and values integrated into the existing culture. A social movement, if it is successful, passes through six distinct phases (Figure 2.6).

The notion that quality has taken on the characteristics of a social movement is pre-

Figure 2.6
Social Movement Model

sented here not as an ivory-tower concept for intellectual types to debate over white wine and Brie. Rather, it is a practical concept that clearly illustrates the power of integration of new ideas and concepts into everyday activities.

Leaders who are not sensitive to the social movement aspects of quality are led by the movement — they do not lead it. They react to changes in the environment rather than create an environment favorable to their organizations. They pursue defensive, internally focused strategies that drive them to improve what they do, rather than pursue offensive strategies that drive them to do what customers require. On the other hand, leaders who consciously or intuitively understand that in dealing with quality they are dealing with a social movement can more easily integrate quality concepts and values into ongoing business practices, drive their quality initiatives with business objectives, and improve the overall capability of their organizations.

The Social Movement Model in Figure 2.6 suggests that if indeed quality is a social movement on a national level, it must pass through all six phases, ultimately arriving (if it is successful) at an Integrative Phase, where its concepts and values become part of the fabric of society. Does the same fate await the quality movement in the context of the corporation? Must a company's quality movement pass through all six phases before there is any payoff?

Figure 2.7 illustrates what we call the Quality Integration Model. Graphically, the

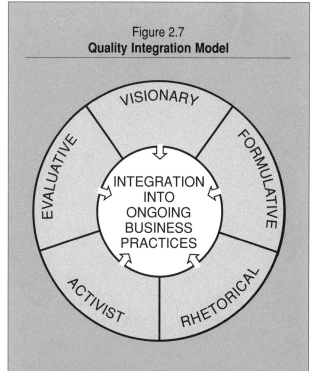

Figure 2.7
Quality Integration Model

Visionary Component: The organization debates, then accepts, that its ultimate objective is creating committed customers. It integrates customer commitment into ongoing business practices.

Formulative Component: The vision of creating committed customers is operationalized by defining a management process that measures and monitors customer relationships.

Rhetorical Component: The process for achieving the operationalized vision is communicated as a call to action to all members of the organization.

Activist Component: The process for achieving the operationalized vision is implemented throughout the organization; behavior change occurs that has a direct impact on customers.

Evaluative Component: Quality activities are measured against their contributions to the organization's defined business objectives.

Integrative Component: Quality objectives and business objectives are one and the same. Creating committed customers is "just the way we do business."

difference between Figure 2.6 and Figure 2.7 is that integration, the ultimate goal of a social movement, has been moved in Figure 2.7 to the center of the circle, touching each of the other components. The symbolic point is that if a leader consciously makes integration the goal of activities that correspond to each component of the model in Figure 2.7, his or her quality initiative will consist not of random quality-related activities, but of quality processes that are linked directly to fundamental business results. The leader and the company will not be dragged sequentially through six *phases* of quality; the leader can manage the six *components* of quality simultaneously in an integrated manner.

How does this discussion relate to the central question of our inquiry — "Is your quality initiative paying off?" As put forth in Chapter 1, the payoff of a quality initiative is directly proportional to the strength of the linkage between quality and ongoing business practices. In this chapter, we have established that the source of that link lies not in internal processes and procedures. The integration of the quality initiative with ongoing business practices, the refocusing of a company externally on its customers, requires that everyone in the company adopt an external focus on customers, *and* that the company has a system in place that facilitates integration of quality with everyday business plans and practices. Defining and driving that system is primarily the role of the company quality leader. It is the quality leader who must initiate the move from an internal, process-oriented view of quality to an external, customer-focused perspective. As we'll see in Chapter 3, defining and driving a management process goes well beyond mere rhetorical support of quality. It involves addressing each of the components of the Quality Integration Model and integrating them into ongoing business practices.

1   David L. Stum. "Customer Satisfaction: Continuing to Build the Service Organization of the Future," a presentation at the 1991 National Communications Forum, Chicago, IL.

2   Theodore Levitt. *The Marketing Imagination.* Copyright 1983, 1986 The Free Press, a division of Macmillan, Inc. New York, NY.

3   *Working For Customers.* Copyright 1983, 1987 Confederation of British Industry, London.

4   *Quality: Executive Priority or Afterthought?* A survey conducted by The Gallup Organization for the American Society for Quality Control. Copyright 1989 American Society for Quality Control, Milwaukee, WI.

5   "Quality: A Job with Many Vacancies," *Quality Progress,* November, 1990.

"Inspirational communication is a plus, but it is only a part of the communication equation. It is the balance of the equation — communicating the call to action and the plan for achieving those objectives — that makes or breaks a quality initiative."

3

A Failure to Communicate

SHIMMERING HEAT WAVES RADIATE OFF THE PAVEMENT OF THE SOUTH Georgia highway. The hot, humid air hangs heavy, like a wet wool shirt. It's too hot for birds. The only sound is the mesmerizing clip clip ... clip clip of thrashing hoes swung in metronome motion by the shirtless chain gang chopping roadside weeds. In the distance, skimming the water-like mirages glistening on the blacktop, a long, black car approaches. It brakes to a stop, showering the prisoners in a flurry of chaff.

Looking every bit the role of the petty tyrant prison warden, an exasperated Strother Martin emerges from the front seat of the car. Two burly guards drag a man from the back seat. Martin scowls at a bloodied but unbowed Paul Newman lying on the ground where the guards disposed of him. Speaking in a slow southern drawl, he delivers the signature line of the movie "Cool Hand Luke": "What we have here," he says, "is a failure to communicate."

Communication, or rather the failure to communicate, is the scapegoat for a growing number of the world's ills, from the failure of personal relationships, to business failures, to disasters like the ill-fated cold-weather launching of the space shuttle Challenger (the result of a failure of communication between project engineers and senior management in charge of the project). Lack of communication is also frequently cited as the reason for failing quality initiatives. Most frequently the blame falls on the company CEO. "He didn't give the program any visible support" or "She just isn't a charismatic leader" are typical of the arrows fired at management when a quality initiative fails.

We believe that such criticism is not justified as frequently as it is used. Charisma creates its own problems (see sidebar "The Charismatic Leader"), and what is seen as lack of visible management support is really the lack of a comprehensive strategy for moving from quality objectives to the integration of those objectives in the ongoing activities of individual employees.

It is during the Rhetorical Component of a quality movement — before the Activist Component — that quality initiatives tend to bog down. The need to communicate quality objectives is a given. Communication is the right "what to do;" the "how to do it" is where companies are failing.

Some companies take a Strother Martin "the flogging will continue until morale improves" approach to communication in support of their quality objectives. They turn up the heat and flog away at employees, shotgunning a variety of communications approaches and methods, hoping employees will finally "get the message." Communication is one-way — management talks and employees listen and obey. As one corporate citizen lamented, "Ours is not to think or ask, ours is but to type and fax."

Other companies take a "whips and whistles" approach to quality communication, promoting slogans over substance. They are typified by a *Wall Street Journal* profile of Fidelity Investments, which in 1988 joined the

The Charismatic Leader

Is charisma the essence of leadership? Researchers David Nadler and Michael Tushman don't believe so. These common problems surround hallowed chieftains:

- Charismatic leaders set unrealistic expectations.
- They create an organizational dependency in which everyone freezes for the next direction from the boss.
- Lower levels of management can't lead well because no vision, exhortation, reward or punishment is meaningful unless it comes from the leader.
- Top management's ability to deal with certain issues is hemmed in by the range of the leader's skills.
- Charismatic leaders create a need for "continuing the magic."

Charismatic leaders can thus turn an organization's employees into a bunch of junkies thirsting for a fix of magic instead of helping them focus on a clear and steady vision.

R. Dilenschneider
A Briefing For Leaders, 1991

ranks of leading American organizations creating executive-level quality positions. Company executives tapped a marketing manager as their quality guru. Immediately she struggled to avoid having her mandate trivialized. The first suggestion from top officials — which she rejected — was "logos, buttons, and mugs." "They wanted to turn it [the quality initiative] into a marketing campaign," she is quoted as saying.[1]

"Rhetoric about excellence is well ahead of reality in ... most companies."

D. Kanter and P. Mirvis
The Cynical Americans, 1991

Having recognized the need to improve quality, companies like the investment firm immediately shift the corporate communications department into high gear, pumping out slogans, designing logos and ordering buttons and mugs printed with the quality message for every employee. When challenged, management at these companies responds that these methods "build quality awareness." Although quality awareness is certainly a *necessary* element of success, equally certain is that quality awareness is not *sufficient* in and of itself to ensure the successful integration of quality into ongoing business practices.

A communication strategy that focuses exclusively on creating awareness creates a state not unlike narcotic addiction. A quality awareness campaign provides an immediate rush and a lot of frenzied activity. To keep the momentum requires another enthusiasm fix — another "Quality Day," another button, a catchier slogan. And like progressive addiction, each successive awareness event must be bigger, better, and more stimulating than the one before just to achieve the same high. An awareness focus for quality communication spawns an endless cycle of activities, each lacking the power to sustain the quality initiative. The quality initiative will stall in the Rhetorical Component of the movement, and quality will never become integrated with the ongoing normal operations of the business unless there is a specific call to action.

Issue a Call to Action

A critical success factor for a quality initiative is a call to action that links quality objectives with new behavior and the new behavior with improved business results. The leader of the quality initiative must not only define the organization's quality objectives (the stuff of awareness), but must also communicate the process —

the specific behavior changes that employees must make — that will move the organization from objectives to measurable results.

Contrast that approach with the more typical quality kickoff approach of launching a quality initiative with a four-star quality spectacular exclusively focused on quality awareness. "Change people's attitudes," the communication people tell the CEO, "and behavior changes will naturally follow." However, psychological research and practical experience dispute that advice.

Although it is true that people can be motivated to change their attitudes, a change of attitude in and of itself cannot ensure *behavior* change — and behavior change is the ultimate objective. As a would-be leader of your company's quality initiative, it is important to your communication effort to recognize that, in general, attitudes are not always the best predictors of behavior. However, some attitudes are better predictors than others, and it is clear that attitudes predict behavior much better when they are linked to action, not just awareness.

For example, when an attitude is formed by direct experience, it usually leads to a change in behavior. Employees may be exhorted to "focus on the customer," but until they have actual, direct experience with customers, the words of the customer focus message are just that — words. It is the attitudes that employees develop while working directly with customers that will ultimately determine their behavior.

Another case where attitude predicts behavior is when expression of the attitude is narrowly focused and associated with a specific behavior. A general attitude expression like "I'm committed to quality" says little about what a person might do if the "quality" at issue would cause his or her organization to slip a development schedule by several months. However, "When quality and schedule collide, we will do what's right by the customer" is a more specific expression. It is not predictive in the sense that it will tell you what action a person will take, but it does indicate the basis or the rationalization the person will use for the course of action ultimately chosen.

At the far end of the spectrum, attitudes are poor predictors when behaviors are codified as habits. A person may express the attitude "Smoking is bad for you," but if that person has a three-pack-a-day habit, his or her attitude does little to predict behavior. If a company habitually makes decisions based on cost and schedule rather than the impact on the customer, a "customer focus" campaign will accomplish little in terms of actual behavior changes.

The point is that all too often, quality initiatives stop short of taking that final step linking quality attitudes to quality behaviors. Trying to demonstrate a "visible commitment to quality," business leaders paint elegant visions of the grand scheme, but they do not tell employees what their *specific* roles are or what actions they should take. Employees are left with a mandate as vague and informative as "do good and avoid evil."

It is reasonable to assume that without a specific call to action, employee attitudes toward quality and the customer are *not* formed from direct experience, not specific to targeted behaviors, and in many cases the politically correct quality platitudes portend attitudes that are contradictory to prevailing habitual behaviors. The CEO's inspiring speech catapults the quality movement into the Rhetorical Component; however, employees leave the quality kickoff meeting charged and ready to go, but with no idea what *specific actions* they should take to participate in the CEO's brave new agenda.

"The message here for both business leaders and quality assurance professionals is that it is naive to assume that employees in any kind of organization will automatically see the logic of quality improvement. The challenge is to design better quality improvement approaches that engage more segments of the work force by appealing to their most powerful personal interests while, at the same time, overcoming the skepticism and apathy that still seem to be part of the fabric of American business."

Conclusion of ASQC/Gallup Study, 1990

The "Be a Tiger" Syndrome

There are a lot of Dinsdales in the business world. Dinsdale is a character in a Jerry Van Amerongen *The Neighborhood* cartoon. One of Van Amerongen's single-panel cartoons illustrates Vice President of Sales Dinsdale at the rostrum wearing a tiger hand puppet. Clutched in the cat's mouth is a sign reading "Be a tiger." A shoe clangs off Dinsdale's head and the caption in Van Amerongen's trademark understatement reads, "Dinsdale underestimates the sophistication of his sales force."

When it comes to quality, one of the biggest mistakes business leaders make is underestimating their audience. Contrary to what many business leaders believe, the employee population today is not poised and pulsating for another burst of executive quality enlightenment. They may not heckle or launch their footwear, but make no

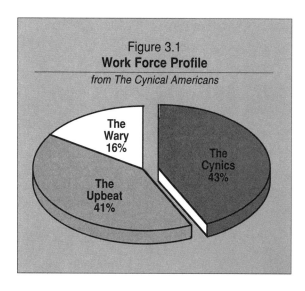

Figure 3.1
Work Force Profile
from The Cynical Americans

The Wary 16%
The Cynics 43%
The Upbeat 41%

mistake about it — when the executive preaches quality, he or she is facing one tough crowd.

In research for their book *The Cynical Americans*, Kanter and Mirvis found that 43 percent of the American working population fit the profile of a cynic (Figure 3.1) — an individual who believes "lying, putting on a false face and doing whatever it takes to make a buck are all part of our basic human nature." Although some of these cynics describe themselves as realists, most are unvarnished cynics, who see selfishness as fundamental to people's character and believe that people inwardly dislike putting themselves out for others. They also believe that the unselfish are taken advantage of and think that people are simply dishonest by nature.

Kanter and Mirvis classified 16 percent of the remaining 57 percent of the work force as wary — those who believe that people are motivated primarily, but not solely, by self-interest. Only 41 percent of the work force in the Kanter/Mirvis study could be described as having an upbeat view of the world.[2]

What is the underlying cause of this cynicism? Kanter and Mirvis explain it as the failure of society to meet the expectations it sets in the minds of the people. The cynical cycle is set up because of unreal expectations reinforced by politicians, the media, school-teachers, success-oriented lecturers and clergy. One is expected to be happy and win. When the inevitable happens and many people do not win, cynics have their built-in explanations, which seldom take into account their own excessive expectations, personal short-comings, or even rotten luck.[3]

To state Kanter and Mirvis's theory in terms of your quality initiative, a communication strategy that focuses on changing attitudes sets up the expectation that things will change for the better. Without an action plan associated with that rhetoric, those expectations can't possibly be met. The result is not just no gain — the result is more scar tissue on the basic cynical attitude prevalent in today's work force. If employees don't see things change, they become even more skeptical.

Communicating a Strategy for Change

So here you are. As a company leader, it is not enough to use your corporate pulpit to preach quality. And every time you do take to the stump for the big "Q," more than half your employee audience is sitting silently cynical of everything you say. And yet, everywhere you turn today, there is a quality "true believer" preaching that the success of a quality initiative depends on the "commitment of the leader" and the leader's "visible support" of the quality initiative. There are more than enough quality terrorists lurking in the corporate woodwork ready and willing to beat management upside the head for a lack of dedication to quality. Can leaders only conclude if they aren't a combination of Knute Rockne, John Kennedy, and Martin Luther King, Jr., then they haven't a prayer of leading their companies to the economic milk and honey of Qualityland?

Many would have you think that is the case. We do not. Inspirational communication is a plus, but it is only a part of the communication equation. It is the balance of the equation — communicating the call to action and the *plan* for achieving quality objectives — that makes or breaks a quality initiative. Providing feedback on performance closes the loop.

Madison Avenue Quality

Winning the 1990 Malcolm Baldrige National Quality Award was a significant accomplishment for General Motors' Cadillac division. The award was acknowledgment of the effort Cadillac made to improve the quality of both its product and its service. Can General Motors be blamed for making the most of that accomplishment? Unfortunately for GM, the answer is "yes."

With all the flourish Madison Avenue could muster, GM touted the Baldrige Award to a somewhat astonished public. While owners of new Cadillacs were aware of improved quality, the general public did not share that view. To make matters worse, GM tended to overstate the implications of the award,[4] which further deteriorated the credibility of their claims.

While GM's rush to publicize "quality" is a visible example of style first/substance later, it is by no means unique. Just as companies tend to resort to buttons, slogans and mugs to promote quality internally, they turn to the advertising department to carry the quality message outside the company and build credibility with customers. Certainly, advertising plays a role in communicating a company's marketplace position, notes marketing consultant Regis McKenna, but advertising can only reinforce a product's position — it cannot create it. Positioning a company or product is no longer just a matter of selecting the appropriate advertising slogan. "It demands a special relationship with the customer and the infrastructure of the marketplace...," notes McKenna. "It is not what you say to your customers, but rather what you do with your customers that creates your industry position."[5]

"A management system is a must. In order to have quality throughout the organization, you must have a guiding system that is followed passionately. Most quality programs fail because they lack either a system or passion, or both."

T. Peters
Thriving On Chaos, 1988

True, great leaders are associated with grand visions and great rhetorical skills. "Ask not what your country can do for you, ask what you can do for your country," is how we remember John Kennedy, not by the arcane details of his tax reform act. Few know the economic theory of Franklin Roosevelt's depression-era recovery programs, yet most people associate FDR with "Fireside Chats" and know that he told Americans that they "have nothing to fear but fear itself." If you need a single example of the power of the Rhetorical Phase of a social movement, imagine how America might be different today if Martin Luther King, Jr., had traded in his "dream" and addressed the masses with the stirring refrain — "I have a strategic plan."[6]

All great leaders have the ability to inspire. But the converse — that the ability to inspire does not in and of itself make a great leader in the world of social or quality reform — is often overlooked. Nor is lack of a charismatic communication style an unconquerable curse. Here we might draw an analogy from the world of sports.

At every level of athletic competition, from amateur leagues to the pros, are coaches who envision themselves as reincarnations of Knute Rockne. Every game is preceded by a "win one for the Gipper" speech that sends the team onto the field on a wave of enthusiasm. However, without a game plan that the team believes in, without each player knowing what he or she must do to contribute to that plan, wins are few and far between. And with each successive failure, the pre-game speech loses a little more credibility.

Successful business leaders, like successful coaches, understand that inspiration can take you only so far. In the case of a quality movement, "awareness" can move a quality initiative to the Rhetorical Component, but not beyond. It can influence what people think about and talk about, but inspiration alone does not provide an outlet for people to actually do something that yields *results*.

Whether you look at business or sports, the business leaders and coaches who consistently come out on top are those who blend inspiration and perspiration and communicate a winning strategy for change. That strategy is two-fold: First, it provides a definitive set of objectives to the organization; second, it communicates concise, easy-to-understand actions that individual members of the organization can take in order to do their part in achieving those objectives. In organizations where a culture change is required, it is the definition of expected behavior changes, not a statement of objectives, that indicates to the organization that the current practices are no longer acceptable and that there will, indeed, be change.

John Kennedy didn't leave people asking what they could do for their country — he created programs like the Peace Corps that provided the opportunities. Franklin Roosevelt didn't leave people to ponder the question of fear — he changed the role of the federal government and created thousands of jobs to involve the people in economic recovery. And when Martin Luther King, Jr., revealed his dream, he and his people took to the streets in a non-violent but rebellious statement that the status quo had changed.

"Momentum comes from a clear vision of what the corporation ought to be, from a well-thought-out strategy to achieve that vision, and from carefully conceived and communicated directions and plans that enable everyone to participate and be publicly accountable in achieving those plans."

M. De Pree
Leadership Is an Art, 1989

A One-Two Punch

Communicating the quality strategy is a two-fold process: First, the leader must provide a clear sense of direction to the organization that focuses the organization externally on the customer. Second, he or she must communicate concise, easily understood actions that employees can immediately take to do their part in moving the organization in the desired direction. The latter requires that you build three key points into your quality message.

First, your message must contain a specific call to action. Second, clearly defined behaviors must be in place that provide an outlet for people to heed the call to action. Third, the action must be directed at specific results so people have a means of measuring their progress.

The best means of realizing the power of the one-two punch of objectives *and* action plan is looking at a case study — a manufacturing company of approximately 2,000 people.

For many years, the company in question enjoyed a market share for its product of approximately 85 percent. However, in a relatively short period of time, two Japanese companies and a second American company aggressively went after its customers. In just two and half years, the company's market share declined to

47 percent. During this dramatic decline, the company remained internally focused, concentrating on reducing internal costs rather than focusing on the marketplace. Instead of looking for ways to grow, it began the process of reducing costs and headcount. Employee morale was low and was having a negative impact on productivity. Business results did not improve, and ultimately the CEO left the company "to pursue other interests."

One of the new CEO's first actions was to rent a very large auditorium and gather all company employees together. He provided employees with a realistic picture of the decline in market share the company had suffered, and further explained that the company was on the verge of losing business with other major customers. The quality of both the company's products and service had been exceeded by the competition.

At this point, however, he departed from what might be the normal course of action. He did not exhort employees to "try harder and do better." He did not lay out a strategic plan that told employees how the company was going to make a comeback. Instead, he showed a single slide with the names of the company's three major competitors.

"What I need you to do," he told employees, "is help me understand how we're going to beat these guys." He asked that each employee sign up for a competitor team. He asked each employee to learn everything they could about the competition. In addition, he asked each employee to interview customers, both existing customers and those lost to the competition. Finally, he asked that in eight weeks each team make a presentation on their findings. Keep in mind that the majority of the employees in this company were hourly, line manufacturing employees.

By this effort, the CEO identified behaviors that would turn the perspective of the organization outward. He did not ask them to identify what was wrong inside the company. He asked them to find out what it took to serve their customers. He empowered employees to understand who their customers were, and what their customer's requirements and expectations were.

During the first week following the presentation, 25 percent of all employees had signed up for a team. By the second week, 60 percent were on board, and by the third week, 85 percent of line employees were out talking to customers and researching the competition.

After eight weeks, the CEO reconvened all employees. On stage were representatives from the three competitor teams. Each team made a presentation on what they had found out — and they found out plenty. From privately held companies they found out gross sales, cost of production and inventory levels. They dissected the competition's product. They presented lists of customer requirements and expectations garnered directly from talking with customers — both current customers and customers who had defected to the competition. Finally, they presented recommendations based on the findings. Following through on his commitment to the process, the CEO empowered employees to carry out those action plans.

In two years, the company's share of the market rose from 47 percent to 77 percent. Although they never reached the 85 percent level again, today their market share has stabilized at 80 percent.

The key to this company's success was that the CEO did not just state the need for change. Instead, he provided a call to action and defined behaviors that motivated employees to get involved and have an impact on their own futures and the future of the company. He committed to an action that visibly showed employees that current business practices were no longer acceptable. Essentially, he took a depressed organization and provided them a clear objective — meeting customer needs and exceeding customer expectations — that got them externally focused and actively working for a cause greater than themselves. He gave employees a simple process that provided identity, purpose and hope.

Communicating a Mechanism for Change

The leader who can provide a message of hope, a message that focuses the individual on something larger than self, will find a receptive audience, despite any underlying cynicism. "The life committed to nothing larger than itself is a meager life indeed," writes Martin Seligman in his book *Learned Optimism*. "Human beings require a context of meaning and hope." Traditionally, that context was found in a belief in the nation, in God, in one's family or in a purpose that transcended an individual's life. However, as

"Quality initiatives fail because they use words, which are the medium by which we communicate and persuade, but not the means by which people change their behavior — and behavior change is fundamental to culture change."

Michael C. Beer
Professor, Harvard
Business School

"To this day, American boardrooms and executive offices are full of senior managers who are all for signing service proclamations and making bold speeches reminding 'those in the trenches' how important customers are, but who can't seem to find the time to actually 'work' the problem of improving bad service."

R. Zemke and D. Schaaf
The Service Edge, 1989

Seligman points out, events during the past 25 years have "so weakened our commitment to larger entities as to leave us almost naked before the ordinary assaults of life."[7]

Notes Seligman, an individual "stripped of the buffering of any commitment to what is larger in life, is a set-up for depression." When individuals face failures they cannot control, they become helpless. Helplessness becomes hopelessness, and may escalate into full-blown depression.[8]

Often, when individuals are depressed, they turn inward. Their thoughts and conversations are self-centered. They feel powerless to change their situation. Without a larger context in which to find meaning, they attribute commonplace failures or disappointments to permanent, pervasive and personal causes. At the extreme, individuals become immobile, both emotionally and physically.

The same sense of depression can overcome an organization when it becomes internally focused. The more a company focuses on internal processes and activities, the more it tends to view commonplace business problems as signs of permanent, pervasive and personal failure. The result of this kind of rumination essentially immobilizes the organization, making it incapable of adapting to changes in the external environment. This state, in which employees feel powerless to influence the company's (and hence their own) future, is already occurring in organizations today. Fewer than half of the employees we surveyed feel they have the freedom to do what it takes to get their jobs done, a paltry 11 percent feels management is responsive to employees, fewer than one-third feel free to make constructive suggestions, and only 37 percent feel their jobs contribute to the success of the business (Figure 3.2).

One method of treating depressed individuals is to change the individual's cognition from an internal to an external perspective. The person's thinking must be reversed. As part of the treatment plan, the therapist may even give the individual a set of structured activities that focuses the individual outside of themselves, often on the needs of others. Serving others provides a purpose that cannot be fulfilled by a preoccupation with self.

Organizations that are depressed require the same treatment — transferring the organization from an internal to an external focus.

The focus must be shifted from internal processes to the external needs of customers. That message is know who your customers are, what requirements they have, what expectations they hold and what you must do to meet those requirements and exceed those expectations. Meaning for the organization comes from the value it creates by serving its customers. Because of the social movement nature of quality, how a leader communicates the quality message — as mere rhetoric or as a clear call to action — can be either the right therapy for a company or the final push into spiraling decline. However, and this is the point of the above discussion, that message alone is not enough. As in the treatment of individual depression, organizational depression must be treated with a structured set of behavior changes that will focus the organization externally.

Defining and communicating the mechanism for change (not just a set of objectives) is the quality leader's primary task during the Rhetorical Component of a quality movement.

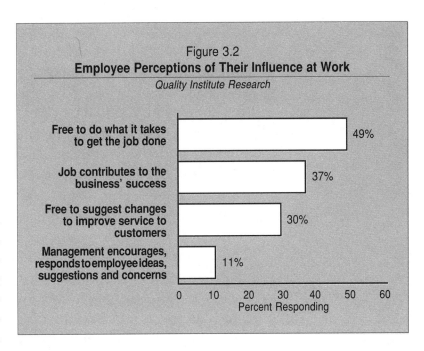

Figure 3.2
Employee Perceptions of Their Influence at Work
Quality Institute Research

I n summary, how do you avoid "the failure to communicate"? Your task — communicating the quality message — is clearly not an easy one. Your audience is tough — at best predominantly cynical, at worst depressed and hopeless, most likely resistant to any implication in your message that they haven't been trying to do quality work all along. Your environment is probably one of constant change. Buttons, mugs, and quality slogans simply won't cut it. Flogging away until quality improves isn't the answer. Pulpit-pounding inspiration ("Be a quality tiger") is likely to get the speaker whacked by a wing tip à la Dinsdale. Is the environment difficult? Yes. Is the task impossible? No.

The CEO in our example faced all of these issues at once, and yet he succeeded in pulling his organization back from the brink. He did it by using communication to integrate quality into the real-time concerns of his employees and into their on-the-job behavior. He succeeded by following a few key precepts. *You can too.*

Your "quality objectives" must focus your organization on knowing your customers and their requirements and expectations (Chapter 4). Those objectives must be linked to a call to action — employees must be charged to do something different. That call must be accompanied by defined behaviors that when implemented provide employees with direct experience with customers that converts attitude into actions (Chapters 5, 6, 7, 8).

All quality communication should support that progression. Recognition for those who respond to customer requirements and expectations should be integrated into normal, ongoing business practices. Every channel of communication should be used to focus employees on their customers. The message should always focus on creating committed customers, not merely satisfied customers (Chapters 9, 10).

Communicating the quality message is not so much a communication project as it is what marketers Ries and Trout call a "selection process."[9] If you want to achieve customer-focused, sustained behavior from your organization, you have to set the stage in a manner that is likely to break through the sheer volume of information transmitted in today's corporate world where a person must check an electronic mailbox, listen to recorded phone messages and check the fax file before finally tackling the in-basket. Where do you think that management memo exhorting employees to "do quality" fits in the employee's hierarchy of things to do?

By providing a method for behavior change, the CEO in the example broke through the everyday clutter that was keeping his employees from talking with their customers. The key was not so much what he said, nor how he said it. The key was empowering employees by providing them with the opportunity to act — to change their behavior in a way that would have an impact on the organization.

[1] "Gurus of Quality Are Gaining Clout," *The Wall Street Journal,* November 27, 1990.

[2] Donald L. Kanter and Philip H. Mirvis. *The Cynical Americans.* Copyright 1989 Jossey-Bass Inc., San Francisco, CA.

[3] Kanter and Mirvis. *Ibid.*

[4] "GM Is Accused On Way It Used Baldrige Award," *The Wall Street Journal,* May 29, 1991.

[5] Regis McKenna. *Relationship Marketing.* Copyright 1993 Regis McKenna. Addison-Wesley, Redding, MA.

[6] Reference to Martin Luther King Jr.'s "strategic plan" used by Sheila Sheinberg in a presentation to the International Association of Business Communicators.

[7] Martin Seligman, Ph.D. *Learned Optimism.* Copyright 1990 Martin E.P. Seligman. Alfred A. Knopf, New York, NY.

[8] Seligman. *Ibid.*

[9] Al Ries and Jack Trout. *Positioning: The Battle for Your Mind.* Copyright 1981, 1986 McGraw-Hill, Inc., New York, NY.

"When presented with the polished and plexiglassed version of 'the mission,' the general employee population ... is totally underwhelmed by the content."

4

The Visionary Focus

WITH THE SUDDENNESS OF A BOLT OF LIGHTENING, THE REVELATION flashes across the consciousness of the corporation — *Our business needs a mission.* Overflowing with the spirit, blinded to all but the revelation, the company president summons his lieutenants. They adjourn to an off-site location. Sequestered from the dual distraction of employees and customers, ensconced in luxurious surroundings, the executive bonding ritual begins. Division heads who just last week were pointing fingers at one another now openly discuss the need for focus and teamwork to achieve objectives of the organization. Innovation, customer focus and employee empowerment take their turns on the agenda. After two or three days of emotional purging, the executive team has drafted (drum roll, please) ... "THE MISSION."

The retreat officially comes to an end when the company president hands a draft of THE MISSION to the PR or corporate communication manager with the charge to "clean it up a little bit" and order plexiglass frames (the modern equivalent of stone tablets) so that each employee can share in the collective wisdom of the visionary few. The "rush to plexiglass"[1] is complete.

Meanwhile, back at the ranch, employees have suffered through three days of anxiety, wondering what the entire executive staff is doing off-site (they are as yet unaware that the executive "staff" has become a "team"). When presented with the polished and plexiglassed version of THE MISSION, the general employee population is relieved that nothing of personal significance — like a layoff — resulted from the meeting, and totally underwhelmed by the content.

Our research has shown that the result of mission statements written under these conditions is a disjointed triad in which THE MISSION, corporate culture and business objectives are out of sync. In our research, we asked employees to indicate the degree to which they understand the mission of their company. We also asked them if they have a clear understanding of how their work relates to the overall plans and objectives outlined by management. Responses indicate that employees are confused by, and lack confidence in, their companies' directions (Figure 4.1).

Defining a Mission — Formulating a Vision

In terms of the Quality Integration Model presented in Chapter 2, the Visionary Component of a quality movement is characterized by a sense of idealism. Within a business organization, this idealism manifests itself as quality for quality's sake. Quality is positioned as the right thing to do, the high road, or the ethical way to run a business. These are all vague statements, however, that carry no specific behavioral changes for employees. In other words, the vision of quality is not yet expressed in operational terms — the definitive characteristic of the Formulative Component of the quality

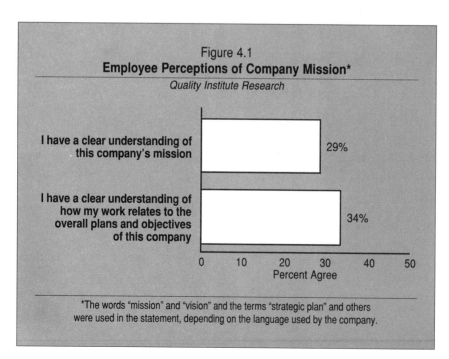

Figure 4.1
Employee Perceptions of Company Mission*
Quality Institute Research

I have a clear understanding of this company's mission — 29%

I have a clear understanding of how my work relates to the overall plans and objectives of this company — 34%

Percent Agree

*The words "mission" and "vision" and the terms "strategic plan" and others were used in the statement, depending on the language used by the company.

movement. A mission statement that does not operationalize vision-ary concepts does not move a company's quality initiative forward. The rush-to-plexiglass syndrome is characteristic of a company whose leadership is caught up in the style of the quality movement. The leaders confuse drafting a mission statement with being visionary.

A mission statement outlines the activities required for business survival and perpetuation of the enterprise. The mission is the back-to-the-basics activity of the business. It must be formulated with the power to energize a company for the long haul.

Writing in *Vanguard Management*, author James O'Toole notes that the "search for meaning" is a vital factor in corporate life. As discussed in the previous chapter, a base or unelevated purpose for a corporation is inadequate:

> *"That is why a company that has as their highest goal (as one that has been recently called excellent proudly proclaims) 'to ensure that our potato chips never get soggy on grocery store counters' cannot be thought of as a great company, even if they do make enormous amounts of money in the short term. It is not that crisp chips are an unworthy business objective ... but, in the long term, the best employees will drift away in search of deeper purpose and more meaningful reasons to devote their lives to a corporation."*[2]

Our research supports O'Toole's conclusion in the sense that companies who are successful in implementing quality initiatives invariably formulate mission statements that are focused externally on their customers rather than internally on their own processes or objectives. The corporate sense of worth is not derived from its efficiency, but on the view held of the corporation by its customers. Companies that have successfully integrated quality into ongoing business practices have focused their corporate mission on the con-cept of meeting customer requirements and exceeding customer expectations.

The Road to Damascus

Formulated or not, every company has a perceived mission. That mission is firmly embedded in the corporate objectives. Employees will strive to fulfill the perceived mission (which is

"Corporate leaders know very well that what seeds the vision are those imperfectly formed images ... about what custom-ers really want and those inarticulate mumblings ... about poor product quality, not crystal-ball gazing in the upper levels of the corpo-rate stratosphere."

J. Kouzes and B. Posner
The Leadership Challenge,
1987

Hearing the Customer's Message

"Quality improvement requires allowing the voice of the customer to guide the firm's activities, telling it where and how to use its resources. To assist in this throughout the firm, marketing must learn to speak the language of other functions and then help those other functions set priorities by determining where customers' need for improvement is greatest. This means, specifically, that researchers cannot be content with merely describing what customers have said but must interpret their answers with more sophistication than ever. For example, it is no longer enough to tell management that customers give the firm a 'B' in its performance. Managers must know exactly what 'A' performance looks like and what actions they must take to create this 'A' performance."

Diane H. Schmalensee, Vice President
American Marketing Association

reinforced by the company's reward structure) regardless of whether it is in agreement with the stated company mission. This lesson is not easily learned, even by the best of companies.

Three years into the Xerox quality program, CEO David Kearns was dissatisfied with the company's progress. The three main objectives of the Xerox program were improved profits as reflected in return on assets (ROA), improved customer satisfaction and improved market share. Each goal was equal to the others in terms of priority and emphasis. Nonetheless, Kearns felt that Xerox's customer satisfaction ratings were not what they should be.

For Kearns, the road to Damascus was a hallway at Xerox headquarters. Walking with Wayland Hicks, executive vice president of marketing and customer operations, he passed one of the many posters listing the three objectives of the Xerox mission. None of the three was numbered. When Kearns expressed his dissatisfaction with customer satisfaction results, Hicks seized the opportunity to point out that although the objectives weren't numbered, the first listed was ROA. Despite the intent, the message to Xerox people was clear — ROA is number one.

As Kearns recalled: "He [Hicks] told me, 'David, in the kids' vernacular, you're going to have to suck it in and say *customer satisfaction is number one.*' Customer satisfaction is now number one. That will drive market share. And the combination of those two things over time will give superior profits."[3]

Our research shows that successful quality initiatives have a single, high-impact focus on meeting customer requirements and exceeding customer expectations. There is no need for corporations to spend the time and energy of its management on formulating a unique mission. Every company ought to have the same mission.

As Harvard Business School professor Theodore Levitt writes, "The purpose of a business is to create and keep a customer."[4] To create and keep customers, a company's mission must be based on four elements expressed as the answers to four questions:

- Who is the customer?
- What are the customer's requirements and expectations?
- What must we do to meet customer requirements?
- What must we do to exceed customer expectations?

The mission of the company cannot be fulfilled unless each employee knows the answers to these four questions. Of course, answers presuppose a system that provides the "how" behind these vital questions: How do we know who our customers are? How do we know their requirements and expectations? How do we translate that knowledge into behaviors that ensure we will meet their requirements and exceed their expectations? Without a system for collecting, disseminating and acting on customer information, a company may well find its quality initiative leading it down the road to the equivalent of the 15-year, defect-free computer chip (Chapter 2).

Vision and Strategy

Once a leader understands that the company's mission is to create and keep customers, the next step is moving the organization toward anticipating and meeting customer needs. Anticipating customer needs is the essence of a visionary, offensive business strategy (not exclusively a quality strategy) that effectively changes the rules of the game. It is a question of outdistancing, not leap-frogging, the competition.

Many corporate leaders, notes author and consultant Kenichi Ohmae, believe a successful strategy is based on creating a sustainable competitive advantage by beating the competition. Traditionally, it is assumed that if your competition is a world-class manufacturer, you must "pick, place, pack and ship" faster and more accurately than they do. If your competition is quick to market, you must be quicker. No matter what it takes, if your competition is good, you must be better. Ohmae, however, debunks that theory. Beating the competition, he says, is not what

"Vision is the essence of leadership. [Knowing where you want to go] requires three things: having a clear vision, articulating it well, and getting your team enthusiastic about sharing it. Above all, any leader must be consistent. As the Bible says, 'no one follows an uncertain trumpet.'"

Father Theodore Hesburgh
Former President
Notre Dame University

The Customer Is Number-One Priority

"The counterpuncher, the reactor, is never world-class, whether in sports, politics or business. Leaders are committed *above all* to customer sensitivity, skillfully and continuously determining and assessing the wants, needs, and possibilities of their current and potential customers.

"This must go far beyond empty rhetoric and breast beating. Uttering pious pronouncements about being 'dedicated to service' and 'committed to excellence' is not enough. In fact, such claims are becoming *counterproductive* in many organizations where action plans are conspicuous by their absence. Rhetoric without demonstrable follow-through creates disillusionment in customers and employees alike."

J. Batten
Tough-Minded Leadership, 1989

strategy should be about. Beating the competition cannot be the first premise of a strategy. The first premise of a strategy is *painstaking attention to the needs of customers.* You test your possible strategies against competitive reality, but you define possible strategies in terms of customers."[5] Take the case of Komatsu and its challenge of heavy-machinery industry-leader — Caterpillar.

A major Caterpillar strength is its dealer organization. Caterpillar has a worldwide dealer network more than 250 strong. It can service any Caterpillar machine anywhere in the world in 24 to 36 hours. How then does newcomer Komatsu compete? With a mission of building a dealer network that better meets the requirements and exceeds the expectations of customers conditioned to Caterpillar's dealer network? Absolutely not. Komatsu focuses on a vision that responds to customer requirements and expectations. It looks at the big picture and anticipates customer needs beyond a dealer network. Komatsu builds machines that require little, if any, service. By changing the rules of the game, Komatsu neutralizes one of Caterpillar's greatest strengths.[6]

The formulation of a visionary strategy and its implementation are two very different things. Our research leads us to the conclusion that effective implementation of a "change the rules" strategy requires that a company understand a basic irony — a visionary strategy is both an extension of and a departure from the status quo. A strategy must be both, or it will not be credible and will fail to motivate.

For example, many companies formulate a quality strategy around their product lines, declaring their intent to become a world-class producer of [name that product line]. At first blush that strategy seems to be the foundation of a visionary outlook. But what if the product line happens to be slide rules? Is being a world-class producer of slide rules visionary? The product-focused quality strategy is an internally focused vision intended to preserve the status quo.

The flip side of the coin is a quality strategy that is based on a radical departure from the status quo. A company with a radical-departure vision typically focuses on becoming the industry leader or number one in the world. Certainly this is a stretch objective, but if an organization is currently number seven in the pecking order and out-financed by the competition, the vision is out of sync with reality in the eyes of employees and customers. Consequently, strategies based on that vision carry little credibility. The visionary is but a voice in the wilderness.

Let's return to the Caterpillar/Komatsu example for a moment. In its strategy to have the best dealer network in the industry, Caterpillar is focusing its activities on its product offering — a machine with superior support. Its vision is preserving the status quo — as long as customers' expectations are that heavy machinery requires frequent service, Caterpillar will remain ahead in the race because Caterpillar provides superior service. However, because Komatsu formulated a strategy based on a radical departure from the status quo — heavy machinery requiring less service — Caterpillar's dealer network risks becoming a market force with "slide rule" potential.

Komatsu is asking customers to cut their support lifeline and trust that the Komatsu machines have the advertised reliability. For this strategy to succeed, it must have some connection to the current status quo; that is, Komatsu employees and customers must have some objective evidence that Komatsu can deliver on its promise. Does Komatsu have in place the quality management process necessary to follow through on its vision? How well Komatsu communicates the answer to this question is the degree to which its vision will be accepted.

The Mission/Vision Hierarchy

The point of the previous discussion is that you need not despair if the business gods have not bestowed upon you a quality vision neatly framed in plexiglass. The roots of the quality vision already exist in your organization and are linked to the generic

"Rather than defining strategies in terms of your competition, then, you define them in terms of customers and how you deliver value to them."

K. Ohmae
Strategy, 1991

Requirements, Expectations and Needs

One of the most confusing quality concepts is the relationship among customer *requirements, expectations* and *needs* and whether "quality" means merely *meeting* customer requirements, expectations and needs, or *exceeding* customer requirements, expectations and needs. This is not a trivial distinction.

In *Quality on Trial,* we are rigorous in stating that an organization must consistently "meet customer requirements, exceed customer expectations, and anticipate customer needs." Those are carefully selected words.

"Requirements" are the tangible aspects of a product or service, the reason a customer makes a purchase. If requirements are compromised, there is no reason for the customer to buy. "Expectations" are the customers' perceptions of how the required product or service should be delivered. Companies should always be looking for ways to exceed those expectations.

Note that "requirements" and "expectations" are not synonymous. Exceeding fundamental requirements has no value for the customer or the supplier. Case in point is the company that produced the computer chip with a 15-year life when customers wanted a chip with a three-year life (Chapter 2). The company exceeded customer requirements (at great cost), but offered the customer no additional value. Exceeding customer expectations, however, holds value for the customer by providing intangibles such as convenience, peace of mind, empathy, and so on.

Consider what happens when you take your automobile in for servicing. You have a basic requirement that the car be serviced properly. You also have some basic expectations — the service manager takes care of you promptly, your car is clean when it is returned to you, and the like. At a minimum,

an auto dealership must meet these requirements and expectations for you to leave a satisfied customer. However, the dealership takes a step toward making you a committed customer if it exceeds your expectations.

How would your opinion of the dealership be altered if, when you brought in your car, the service manager referred to the work the shop performed on your car the last time you came in (maybe six months ago), and asked if you were still satisfied with that work (empathy with your problem)? Or what if a mechanic (not a "runner") brought your car up after service and explained exactly what was done (assurance that the work was done right)? Or a "loaner" car was provided free of charge (added convenience)? Would this dealership have exceeded your expectations?

Note that the dealership must still meet the basic requirement of properly servicing your car. Exceeding expectations never takes the place of meeting basic customer requirements.

"Needs" are tangibles of product or service that are not currently available to customers from any source, but that if available, would quickly become requirements. Again, the automotive world provides some great examples. Many items that are standard fixtures on automobiles today — everything from windshield wipers to rear window defoggers — were at one time unique responses to customer needs *that proved to have value for customers.*

The point to take home is that it is impossible to effectively anticipate needs without developing relationships with customers in which information and ideas are freely and frankly exchanged. When a supplier has proven that it can consistently anticipate the needs of its customers, it has also demonstrated a value to its customer that goes beyond its product or service. It has created a committed customer.

mission of meeting customer requirements and exceeding customer expectations. If you are actively pursuing a customer-focused mission — actively pursuing answers to the questions "Who is the customer and what are the customer's requirements?" and "What must we do to exceed customer expectations?" — the leap to a visionary strategy that expands beyond meeting customer requirements and exceeding their expectations is shorter than you might imagine. Definition of a visionary strategy addresses the questions:

- What will be the customer's future needs?
- What must we do to meet those future needs?
- What must we do to gain customer commitment?

Figure 4.2 puts the Mission/Vision Hierarchy in a graphical perspective. The bottom four layers in the hierarchy are the basis of a company's mission and deal with understanding customer requirements and expectations. The top three layers are visionary and deal with meeting customers' needs. At the very peak of the hierarchy is "customer commitment." Commitment is the end-state goal of a relationship; customer commitment is the end-state goal of a business entity.

By definition a hierarchy is clearly a progression; the bottom layers lay a foundation for the upper layers. Before customer expectations can be exceeded, they must be met. Before requirements can be met, they must be identified. And before customer requirements can be identified, businesses must know *who* their customers are.

From a management perspective, the implication is that the hierarchy cannot be entered anywhere but at the bottom. Companies cannot focus a quality initiative on anticipating customer needs if they are not currently meeting customer requirements. Companies can't exceed their customer expectations if they don't have an understanding of the customer's current requirements. At the extreme, organizations can't focus on customer commitment if they don't know who the customer is. It is equally clear that movement up the hierarchy without a means of evaluating progress at each level is impossible. The operative word is "evaluating." You can't assume that you intuitively know.

"The impetus for instituting quality programs is often reactive — to offset a negative image and a loss of customers to competition — rather than innovative or preventive.

"Among those who acknowledged an unfavorable image of American products, 61% cited a lack of management foresight about quality's importance to customers."

Conclusion of 1991 SPSS, Inc. quality study

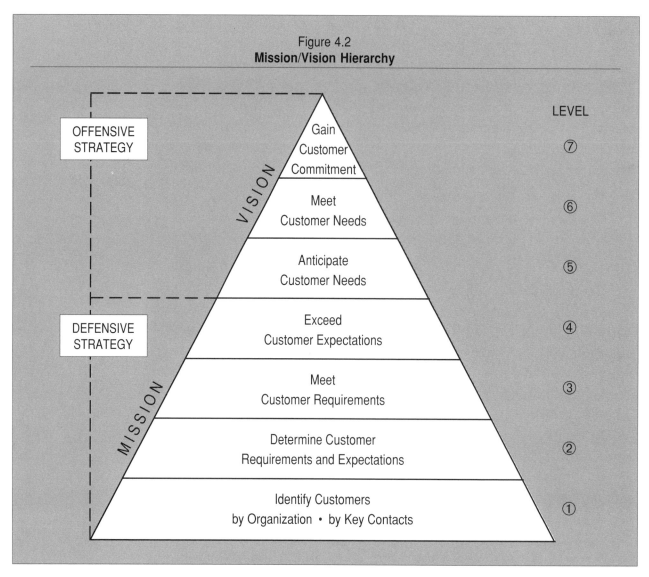

Figure 4.2
Mission/Vision Hierarchy

Leaders who have a customer management system in place that addresses the four questions that drive the mission of the corporation are in the best position to formulate, communicate and implement a vision that will gain customer commitment.

In summary, we have found that the basis of a successful quality initiative is a visionary strategy based on an external customer focus. Meeting customer requirements and exceeding customer expectations is the mission of everyone in the company. It is a requirement to be competitive. However, if we heed the wisdom of

Sun Tzu, who around 500 B.C. noted that the smartest strategy in war is the one that allows you to achieve your objectives without having to fight,[7] it becomes obvious we must outdistance our competition not by challenging its strengths, but by changing the rules of our industry in a manner that cannot be easily replicated by the competition.

Gaining such an advantage does not require a technological breakthrough of Edisonian proportions. We submit that the means to achieve such advantage lies within the capability of most organizations today. That advantage is in having a customer management system in place that provides opportunities for everyone in the organization to continuously improve the quality of the *relationship* they have with their customers.

The Quality Review process has proven to be an effective customer management system (see Chapter 9). It provides the basic information and behavioral model for implementing a customer relationship strategy. It involves line and staff employees in determining who their customers are, and it formulates a company's quality initiative in terms of meeting customer requirements and exceeding customer expectations — the external perspective missing from so many quality initiatives.

But the Quality Review process also contributes to the more visionary objective of anticipating customer needs and gaining customer commitment. The primary intent of the Quality Review process is to establish a formal management system (an equal of the financial management system, the manufacturing control system and other key business systems) for measuring, monitoring, creating and maintaining relationships with customers.

---

[1] "Rush to plexiglass" is a phrase from James E. Ericson, president of Training Services Corporation, Minnetonka, MN.

[2] James O'Toole. *Vanguard Management: Redesigning the Corporate Future.* Copyright 1985 The Berkley Publishing Group, New York, NY.

[3] "The Great Expectations of David Kearns," *Industry Week,* June 17, 1991.

[4] Theodore Levitt. *The Marketing Imagination.* Copyright 1983, 1986 The Free Press, a division of Macmillian, Inc., New York, NY.

[5] Kenichi Ohmae. "Getting Back to Strategy," *Harvard Business Review,* November/December, 1988.

[6] Michel (Mike) Robert. "Changing the Rules of Play," *The Strategist,* Volume 3, No. 1. Copyright 1991 Decision Process International, Westport, CT.

[7] Sun Tzu (translated by Samuel B. Griffith). *The Art of War.* Copyright 1963, 1971 Oxford University Press, New York, NY.

"What's missing from the Ferris wheel viewpoint ... is the ground-level perspective of the relationship between the organization and its customers."

5
Moving Beyond
Customer Satisfaction

"IF YOU EVER FEEL SO FAR AWAY FROM YOUR CUSTOMERS THAT YOU CANNOT identify with them, get down off the Ferris wheel."[1] That's the advice of a Confederation of British Industry booklet entitled *Working for Customers*. The Ferris wheel allusion is to a scene from the British film "The Third Man." In the scene, villain Harry Lime is trying to justify his drug dealing to the film's protagonist, Holly Martins. From their vantage point high on a Ferris wheel, Lime points at the pinhead humans on the pavement below. Surely, for a fortune, he asks Martins, he wouldn't care if one of those dots stopped moving?

Later in the film, Martins is forced to visit a children's hospital and meet Lime's victims. The personal confrontation with the aftermath of Lime's business is a much more shocking and intimate perspective than Martins had from the top of the Ferris wheel. We submit that many corporations today are looking at their customers from the tops of Ferris wheels. They have a lot of data that are dots

"Most market research is aimed at prospects — not customers — to determine the market acceptance of new products or services. Furthermore, research of any kind is seldom used by managers to make customer-focused decisions, let alone incorporated into product and service development, production, and marketing (another secret of the Japanese 'miracle')."

J. Koob Cannie
Keeping Customers for Life,
1991

on charts — market segmentation graphs, market research studies and customer satisfaction surveys. Charts and graphs summarize data at a very high "pinhead" level. By the time these data are translated into action plans (if they ever are), they have been so diced, sliced and chopped by the analytical blender that resultant actions bear only accidental resemblance to the specific needs and expectations of any individual customer.

What's missing from the Ferris wheel viewpoint of macro-level research is the ground-level perspective of the relationship between the organization and its customers. Action plans based on traditional Ferris wheel methods are no guarantee of customer loyalty or competitive advantage. The ultimate survival of any enterprise depends on how effectively it establishes long-term relationships with its customers. Hence, companies must move beyond customer satisfaction toward a customer management system that measures their customers' level of commitment to the organization. The system must also encourage behavior that integrates relationship management into the ongoing business practices of the company.

Moving Toward Customer Commitment

Today, the proverbial "everyone" is measuring customer satisfaction. Buy a car, and two weeks later the little "How Did We Do?" card (or 12-page customer satisfaction questionnaire) arrives in the mail. Eat out — there's a "We Really Care" slip tucked between the salt and pepper asking for your input. In your hotel room, there in the room service menu between appetizers and entrees, is the 3" × 5" plea "Tell Us How We're Doing." And back at the office, chances are good that customer satisfaction surveys from suppliers, both internal and external, have added another inch or so to the annual height of your in-basket.

And what becomes of all your checks, ticks and "additional comments in the space provided"? From our experience, we can tell you that in the majority of cases the data is tabulated, thoroughly analyzed, aggregated and disaggregated, sorted and normalized. Reports are made to management, astonishment is registered, plans are made and ... mostly nothing happens. Why not?

Remember the "telephone" game you played as a child. The kid at one end of the line whispered a sentence that was passed from child to child, distorted with each exchange, until the last one in line, between bursts of laughter, tried to spit out an almost incomprehensible string of words. "Betty bakes with bitter butter" became "Bobby busts his bottom buttons."

What do you think happens to customer satisfaction data collected by a third party and then passed up, down, around and through a business hierarchy? What is the reaction of employees who are the last to receive the information and then are expected to act on it? What often happens is the company frantically sets about fixing Bobby's bottom buttons when the real problem is Betty's bitter butter.

But let's assume that the data from traditional customer satisfaction measures makes it up and down the management chain and maintains a reasonable amount of integrity. What is happening while all the analyzing is going on? Dissatisfied customers are still dissatisfied, problems causing the dissatisfaction are still occurring, and the business environment continues to change. By the time the cycle of survey, analyze, report, plan and implement is complete — a time frame that may be a year or longer — many customers who indicated they were dissatisfied are no longer customers, and the solutions proposed as a result of the survey likely no longer fit the changed environment in which management insists they be implemented.

If you find it hard to come to grips with that view of traditional research, believe us, we understand. Part of the services we provided to major corporations over the past 20 years were traditional customer satisfaction surveys and market research. However, our experience left us less than satisfied with the results. We were the disinterested, objective researchers. We were the third party brought in because we could distance ourselves from the customer and

Satisfaction vs. Commitment

"Satisfied customers feel good as long as their needs are fulfilled; committed customers look beyond short-term pleasures and develop an allegiance to the firm. Satisfied customers are pleased, humored and fulfilled; committed customers are dedicated and faithful. Satisfied customers remain independent from the firm; committed customers become interdependent with the firm through shared resources and values. The totally satisfied customer says, 'My needs have been assessed and met, so I feel good about dealing with the firm'; the totally committed customer says, 'We have developed interdependencies, shared values, and strategies to the extent that our separate needs can best be met through long-term devotion and loyalty to each other.'"

D. Ulrich
Sloan Management Review, Summer, 1989

provide objectivity. And we did a darn good job. The only problem was that we weren't really helping our customers get any closer to their customers. The best we could do was tell them how far away, to a couple of places east of the decimal point, they were from their customers. Our focus was exclusively on the general factors typically associated with customer satisfaction. We were ignoring the behaviors that drive an organization to strengthen its relationships with customers, building toward customer commitment.

Tom Peters and Nancy Austin have quantified a similar experience. They, too, concluded that although executives talk a good game of "customer first," the majority do not have initiatives in place that support that commitment. Peters and Austin surveyed 172 executives, 134 of whom are presidents of companies. Unanimously, the executives ranked long-term customer relationships as the number-one priority in their companies. Yet all 172 executives admitted that they had not assigned accountability for measuring and monitoring these relationships.[2]

The clear conclusion of a 1989 study by Learning International (Figure 5.1) is that the competitive race of the 1990s will be won by the organizations that understand what matters most — and least — to their customers. The study identified and ranked by percentage of impact the six critical factors that contribute to overall customer satisfaction.[3]

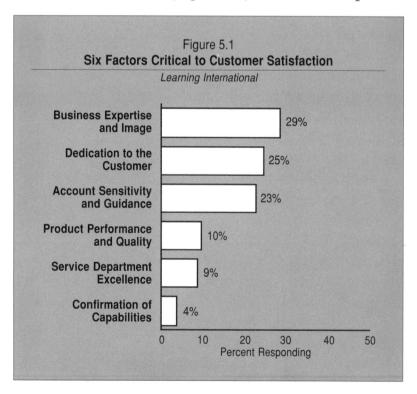

Figure 5.1
Six Factors Critical to Customer Satisfaction
Learning International

This research supports a get-down-off-the-Ferris-wheel approach to the customer/supplier relationship. The tangible factors of product performance and quality and service department excellence — the things that are traditionally measured to death under the rubric of "customer satisfaction" — are not most important to customers. The most important factors to customers are the intangible factors of business expertise and image, dedication to the customer, and account sensitivity and guidance, which are seldom measured or actively addressed.

Normative data in the Quality Institute database (gathered as part of our research and through customer interviews conducted by our client organizations) supports a similar conclusion. Our research ranks intangible factors — a partnership approach, accessibility, responsiveness, reliability and competence of personnel — as most important to customers. These factors even outdistanced those of delivery, price and technical support.

Customer satisfaction measurement is but the ante to play the game. It clearly does not guarantee winning the pot. Customer satisfaction is not the same as, nor is it a guarantee of, customer commitment. Our research shows that more than 40 percent of customers who claimed to be satisfied switched suppliers without looking back. Further, consider the following data:

- 89 percent of people who owned cars from a certain manufacturer said they were very satisfied with their automobiles.

- More than 67 percent stated that they intended to purchase another car from that manufacturer.

- Fewer than 20 percent actually purchased another car from that manufacturer.

Satisfaction is but a short-term state achieved by meeting customers' immediate needs. Customers remain satisfied only as long as they perceive that a supplier is providing value. If no relationship between customer and supplier exists beyond the payment for goods and services provided, then once the customer's immediate needs can be better met by some other supplier, he or she makes the switch — regardless of previously stated intentions.

On the other hand, commitment is a long-term state where shared experiences with a supplier provide customers with a historical perspective on short-term occurrences. Committed customers gain equal, if not greater, value from their relationships with their suppliers. Short-term benefits offered by competitors, like lower price or more bells and whistles, do not disrupt the shared history

"Satisfaction is, as it should be, mute. Its prior presence is affirmed only by its subsequent absence. And that's dangerous, because the customer will be aware only of failure, of dissatisfaction, not of success or satisfaction. That makes him terribly susceptible to the blandishments of competitive sellers."

T. Levitt
The Marketing Imagination,
1986

of the customer/supplier relationship. If greater short-term value exists for the customer outside the current relationship, the first impulse is to work things out within the relationship, not switch suppliers. Consider the case of the "black hole."

A purchasing agent with a question about one of his supplier's products called the headquarters of the supplier (a Fortune 500 company and a Baldrige Award winner). After a lengthy series of transfers, he was routed back to the person who first answered the phone, but who couldn't answer the question. Frustrated, the agent went to his supervisor and declared that he was "never again going to deal with that incompetent supplier."

"That's strange," replied the supervisor, who had worked with the supplier for many years, "that's not the way they do business. Try again."

On the second try, the purchasing agent was correctly routed and received detailed information. Ultimately he made the purchase. Had it not been for the relationship between the supplier and the customer company, the supplier would never have gotten a second chance. A customer would have been lost.

"Leaders need to foster environments and work processes within which people can develop high-quality relationships — relationships with each other, relationships with the group with which we work, relationships with our clients and customers."

M. De Pree
Leadership Is an Art,
1989

In this case, the customer relationship was clearly an asset for the supplier. In fact, it can be argued that the relationship between customer and supplier is the supplier's only true competitive edge. Think about it. Price can always be undercut. A new technology emerges, and virtually overnight, your product-based edge is wiped out. You can have the right product and be satisfying customers today, but what about their needs tomorrow? But no competitor, no matter how aggressively it prices or how advanced its technology, can duplicate your history with your customers.

We see the current business environment as a ripe opportunity to move beyond the traditional focus on customer satisfaction — an opportunity to gain competitive advantage through systematic management of customer relationships.

The Foundation for Relationship Management

Relationship management can mean any conscientious effort to direct the development of ongoing interactions with customers. While there is widespread recognition of the importance of relationship management, there is equally widespread reluctance to tackle it.

If executives recognize the value of moving their organizations up the Mission/Vision Hierarchy from satisfied customers to committed customers (Chapter 4), the obvious question is, "Why aren't there more companies with systematic procedures in place for measuring, monitoring and managing customer relationships?" If creating and maintaining a customer is the undisputed objective of every business, why the hesitancy to partner with the most important assets a company will ever have — its customers?

People have a belief that relationships are somewhat mystical — that examining them too much will somehow jinx the relationship. Analyzing a personal relationship has all the romance and appeal of a pre-nuptial agreement. People are reluctant to apply a scientific or process approach out of fear that such activity diminishes the "magic" of the relationship. Despite the potential benefits of a greater understanding of what makes people tick, notes psychologist R.A.

Leaders Serve Customers

"Most managers focus on the collective groups of customers that make up a *market*, thereby adopting a relatively impersonal attitude toward the marketplace. As with many other aspects of business, the manager's mind thinks about markets in fairly analytical and quantitative terms, and is comfortable with such terms as *market share, market penetration, market segments, market growth, market life cycle,* and so on.

"The leader, in contrast, prefers a much more qualitative and human view of the marketplace and likes to think of a market in terms of individual customers. The leader's soul takes into account the human side of customers — the feelings, emotions, needs, and wants that cause customers to purchase a product or service."

C. Hickman
Mind of a Manager, Soul of a Leader, 1990

Hinde, people "regard the term 'science' as simply inappropriate in this context [the study of relationships] because of the complexity and intangible nature of human interpersonal relationships." And yet, he argues, "it is because the issues [of a relationship] are too close to us to be seen clearly that such an [scientific] approach is necessary."[4]

Hinde is among leading social psychologists who have identified a number of behaviors that are indicative of committed relationships. These behaviors are as follows:

Interactions between individuals are frequent, occur over a long period of time and occur in a variety of settings.

Customer/supplier relationships usually start out as a single point of contact with a single purpose — making a sale. However, an effective relationship strategy requires that the supplier organization

"Customer satisfaction is about anticipating what the market will want tomorrow, next month, and next year, and being in a position to provide it. It is achieved through exploration, not confrontation, facts and data, not intuition ... in short, it is very much a contact sport ... The missing link between corporations and the customer is the human touch."

C. Westland
Quality: The Myth and the Magic,
1990

expand its contacts to include all levels of the customer organization. For example, the supplier's shipping personnel establish effective relationships with the customer's receiving personnel. The supplier's marketing and operations personnel establish effective relationships with multiple levels of personnel in the customer's organization — decision influencers as well as decision makers. These contacts ensure ongoing commitment throughout the customer organization and bring the voice of the customer into the supplier organization at many levels.

Individuals are open to each other and are willing to disclose intimate information.

An effective relationship management strategy starts with the objective of developing "intimacy" with the customer.

Citing business and proprietary reasons, few companies are willing to allow customers access to organizational activities: "We can't let customers have access to our cost structure" or "We can't let customers know there's a performance problem with the new product." We can't let customers "this" and we can't let customers "that," all for the sake of the integrity of the business. At least that's the argument; is it valid?

We submit it isn't. There is a perceived risk only because the customer/supplier relationship is not strong enough to withstand intimacy. Think about one of your close personal relationships — there are things you would discuss with a partner of 15 years that you certainly would not discuss on a first date.

Effective relationship management allows for an open exchange of ideas focused on those business issues that are key to the success of both the customer and the supplier.

Individuals develop effective communication systems, and are willing to provide positive and negative feedback about each other's performance.

In the context of a customer/supplier relationship, effective relationship management ensures that a structured system encourages exchange of both positive and negative feedback in a timely manner. This approach shuns reactive, ad hoc contact in favor of active seeking of customer input. The customer management system supporting that activity must provide training on receiving customer feedback, questioning and listening effectively, and resolving conflict. The system must also include a uniform, easy-to-understand methodology for collecting and responding to customer information. Individuals collect information that makes it possible to anticipate each other's needs.

As we stated in Chapter 4, anticipating customer needs is essential to building customer commitment. As fault-free products become the norm, businesses must seek a higher level of relationship with customers — a relationship that goes beyond meeting customer requirements (satisfaction) to anticipating customer needs (commitment).

We can draw a parallel here between the way some businesses approach their customers and the way some individuals approach the institution of marriage. Some people are in love with the idea of being married. They act the way they think married people should act. They do "married things" and completely lose sight of the real needs of their partners. Likewise, business people are usually in love with the idea of being in business. They do what they think business people should do. They do "oil company things," "telephone things," "computer things," and other "business things," and completely lose sight of the real needs of customers.

Commitment is a two-way street. If businesses want committed customers, they must find ways to anticipate what their customers will need from them. Without a relationship management strategy that involves the supplier organization with the customer organization beyond the point of sale, the ability to anticipate needs is severely curtailed.

Individuals work together toward common goals.

A clergyman we know says that the strength of a marriage is seldom indicated by the intensity with which a couple stares into each other's eyes. A more accurate indicator is whether they are looking in the same direction. The same logic holds at the organizational level. Is the supplier goal compatible with that of the customer? A strong relationship cannot be maintained in the face of conflicting goals.

"Selling a product is not a matter of using a 'bag of tricks.' It is the careful attention to the customer's business and the development of solutions to problems that impede his success. Furthermore, in our industry, it includes the ability to partner with our customers to create new sources of revenue for them that will allow them to compete in their own arenas."

Gil Amelio
President
National Semiconductor

Effective relationship management does not stop with information collection. Employees at all levels must develop a clear understanding of their customers' goals and ongoing business requirements. They must follow a process that aligns their behaviors with the needs and expectations of their customers.

Individuals continuously increase their investment in their relationships.

Effective relationship management ensures that customers are an integral part of the company's strategic planning process, including involvement in product design, manufacturing, engineering and service. Likewise, individuals in the supplier's oganization must invest the time and resources necessary to understand the customer's current requirements and future needs.

Individuals perceive that their relationship is unique and irreplaceable. They feel that their individual interests are best served by maintaining a close partnership.

Effective relationship management clearly demonstrates to customers that their individual business interests are best served by a partnership with the supplier. Suppliers who effectively partner with their customers have the potential to gain a "preferred" status. Likewise, customers who partner with their suppliers can have a significant impact on the quality of goods and services bought by their organization.

Relationship Management and Competitive Advantage

Unfortunately, most organizations have no relationship management strategy that effectively focuses on creating committed customers. Of the organizations we have found that do recognize the need to manage customer relationships, few have any kind of system that drives the behaviors described above. Relationship management remains a theoretical discussion.

Competitive advantage is inherent in effective relationship management in three key ways. First, relationship management drives people within an organization to view their customers as individuals rather than homogenized dots of data — as people with varied needs, both overt and below the surface, tangible and intangible, rational, emotional and social.[5]

Second, relationship management requires ongoing dialogue with customers in order to continuously understand changing requirements and expectations. This interaction increases understanding of the other's goals and needs.

Third, relationship management helps an organization anticipate future customer needs that go beyond product or service. It focuses a company's energy on discovering, creating, arousing and responding to customer needs.[6]

Relationships are true company assets. Like other assets, they can appreciate or depreciate. "Their maintenance and enhancement," says Theodore Levitt, "depend not so much on good manners, public relations, tact, and charm as they do on making an investment in good relationship management."[7] The building of committed customer relationships requires the systematic collection and dissemination of customer information.

The Missing Link

In addition to creating committed customers, most organizations have another vision — to create a community of employees committed to achieving the organization's objectives, which in turn motivates employees to *actively* contribute to the organization's success. Employees are the crucial link between company and customer. Unfortunately, they are frequently the missing link. This is not a unique insight. Ever since some anonymous copywriter coined the phrase "people are our most important asset," organizations have been struggling to add substance to those words. The task is not trivial.

In the era of downsizing, rightsizing, smartsizing and reorganizing, the business world's "most important assets" are feeling more used and abused than infused with company loyalty. Employees are often more concerned with their own survival than with contributing anything to the company beyond the bounds of their 9-to-5 responsibilities. In a time when business magazines run articles asking whatever happened to company loyalty, companies are frantically striving to regain the trust of employees and tap their expertise to improve organizational effectiveness. And therein lies a problem.

"People are being perceived more and more as costs to be reduced, as opposed to long-term assets to be grown and developed. And once you begin to see people as costs, you begin to treat them more as things than as humans — as throw-away employees. In other words, employees are like other things in the production equation; and once the thing is used up, you replace it with another."

David M. Noer,
Vice President for Training and Education,
Center for Creative Leadership

*"Organizations
frequently set up
listening programs to
encourage employees
to share their ideas
but the systems do
not always provide
for timely feedback.
Employees get answers,
often inappropriate,
months after the
question was asked.
By that time, it is no
longer an issue or
employees have lost
trust in the system."*

Personnel Administrator
September, 1988

Companies have gone about trying to gain the commitment *and* tap the expertise of their employees, treating trust and participation as two separate issues. They fail to recognize a basic fact of human nature — trust results *from* participation; they go hand-in-hand. Gaining employee trust and generating employee involvement in the organization are both functions of the relationship a company has with its employees.

How do companies try to foster this relationship? Typically, they survey their employees to death trying to uncover their satisfaction or dissatisfaction with everything from the company benefits package to the CEO's management style. A paper-and-pencil survey is administered. Employees anonymously blacken circles with a #2 pencil to register their attitudes toward the organization. Results are aggregated, charts and graphs are reviewed, Corporate Communication publishes a brochure summarizing survey highlights and lowlights, and management scratches its collective head about how to respond to the dizzying array of employee concerns that don't change much from survey to survey. And while all this surveying is taking place, the Facilities crew is busy tacking up suggestion boxes around the company begging employees for their cost-saving, waste-reducing suggestions, "enticing" them with incentives ranging from a thermos bottle bearing the company logo to giant sweepstakes offering exotic travel. Behind each submitted idea lurks a set of "rules" that reads like a mortgage contract as to how each idea will be evaluated and the employee rewarded.

What's missing is any attempt to integrate employee trust and participation into the day-to-day process of the business. Employees are required to "put down their tools" and attend morale-building events, or "step outside the box" and submit ideas to the company suggestion program. This method encourages management to view employees from a distance — from the top of the Ferris wheel, where flesh-and-blood people are seen only as little black dots on a chart. What message does this method send? What are the consequences of this Ferris wheel view?

When we look at employee surveys and suggestion programs, we see clearly that they are not effective methods for building strong employee/management relationships. Using Hinde's analysis as a guide, we find that these programs do not lend themselves to sharing intimate information, do not provide *timely* communication between individuals, do not foster a sense of working toward common goals, and do not require much of a personal investment in the relationship on behalf of either management or employee. Bottom line, they contribute only marginally to increasing the effectiveness and efficiency of the organization.

As professional researchers, we recognize the value of survey data, but have come to realize that surveys and suggestion boxes in and of themselves do not alter behavior within the organization. These traditional programs do little (if anything) to focus employees on the customer, improve employee morale, open channels of communication, build employee confidence in the organization, or tap into employee expertise. In fact, it can be argued that instead of bringing companies closer to their mission of creating committed customers, traditional methods used to involve employees in improvement initiatives actually perpetuate the notion that employees *cannot* be open with management. They often do more to alienate employees than they do to ally them. If there were an atmosphere of trust, why the emphasis on anonymity? If management really wants employees' suggestions, why don't they ask — really ask — instead of merely putting up suggestion boxes?

Competitiveness in today's environment demands that organizations get down off the Ferris wheel and get a ground-level perspective of their customers and employees. Macro-level data may have their place as a tracking mechanism, and yet collecting and acting on such data is at best a reactive and ineffective way of changing an organization's behavior. Macro-level data measure how well an organization is managing customer and employee relationships; they do not tell the organization how to manage those relationships better.

"Years ago, when corporations still wanted employees who did only what they were told, employee surveys and walk-around management were appropriate and effective tools. They still can produce useful information about routine issues like cafeteria service and parking privileges, and they can still generate valuable quantitative data in support of programs like total quality management. What they do not do is get people to reflect on their work and behavior. They do not encourage individual accountability."

Harvard Business Review
July-August, 1994

In Chapter 6, we'll describe how to manage *customer* relationships better. We'll discuss the Quality Review process, a system for managing long-term external and internal customer relationships. This process facilitates the building of customer relationships by formally integrating into ongoing business practices those behaviors that contribute to committed relationships.

Realizing that employees who are committed to the company's vision are the key link to committed customers, in Chapter 7 we'll take a look at TALK². TALK² is a process that applies the principle of face-to-face interaction between employees and management, and creates a system for focusing employees on customer issues, building employee trust, and effectively seeking their suggestions for improving the organization.

The principles of the Quality Review process can also be used to manage an organization's relationship with its suppliers — the topic of Chapter 8.

[1]   *Working for Customers.* Copyright 1983, 1987 Confederation of British Industry, London.

[2]   Tom Peters and Nancy Austin. *A Passion for Excellence: A Leadership Difference.* Copyright 1985 Warner Books, New York, NY.

[3]   *Profiles in Customer Loyalty.* Copyright 1989 Learning International, Inc., Stanford, CT.

[4]   R.A. Hinde. "Interpersonal Relationships — In Quest of a Science." *Psychological Medicine*, Volume 8, 1978.

[5]   *Working for Customers.* Copyright 1983, 1987 Confederation of British Industry, London.

[6]   Theodore Levitt. "Marketing Myopia," *Harvard Business Review,* September/October, 1975.

[7]   Theodore Levitt. "After the Sale Is Over...," *Harvard Business Review,* September/October, 1983.

"Companies simply do not have the right tools to gather or use customer information, nor do they have a methodology for measuring and monitoring customer relationships."

6

The Quality Review Process: Managing Customer Relationships

T HE SUN NEVER SET ON THE CORPORATION'S QUALITY INITIATIVE. THE CEO saw to that. Using the corporate jet as his winged chariot, he raced the sun around the world spreading the quality message. Tropical breezes blew off the azure waters of the Pacific as he rallied the Far East region employees around the banner of total customer satisfaction. "The very survival of our company," he solemnly told them, "depends on everyone's commitment to our customers."

Amid the excitement of the consolidation of a continental economy, the CEO applauded the efforts of the European region employees for their active support of the quality initiative. "Quality is vital to every stakeholder in the company," he intoned. "Shareholders, customers, suppliers, employees, and other constituents depend on the corporation to provide jobs, pay taxes, and make contributions that benefit the countries and communities we serve."

Back home in heartland America, he made a passionate plea to the employees of the domestic Midwest region. "Everyone," he emphasized, "must give their full cooperation and effort to serving their customers by doing their jobs right the first time and every time."

With waving wheat fields fading in the distance, the corporate jet winging its way toward company headquarters, and his mission completed, the CEO nestled all snug in his chair while visions of quality danced in the air. Had he not adhered to the tenets of the quality gurus? Had he not personally sown the seeds of a quality harvest like a Crosbyesque version of Johnny Appleseed? Had he not preached the right message? Had he not very visibly demonstrated his and the corporation's total commitment to quality? He had, indeed.

But imagine the CEO's reaction when presented with the results of a worldwide survey we conducted of his corporation's employees. The survey, conducted within six weeks of his circuit ride, asked employees a series of questions about their jobs, their understanding of the company's mission, and about their overall satisfaction with the corporation.

Almost two-thirds (64 percent) of employees did not know who their customers were and 72 percent did not know their customers' requirements. Almost two-thirds (63 percent) said the quality of the products and services they received from their internal suppliers (other departments within the corporation) did not meet their requirements. More than one-half of employees (57 percent) did not understand the corporation's mission. And 83 percent did not know what to do to positively impact the mission (Figure 6.1).

What had gone wrong?

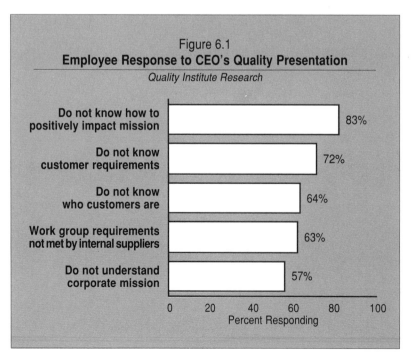

Figure 6.1
Employee Response to CEO's Quality Presentation
Quality Institute Research

Response	Percent Responding
Do not know how to positively impact mission	83%
Do not know customer requirements	72%
Do not know who customers are	64%
Work group requirements not met by internal suppliers	63%
Do not understand corporate mission	57%

The CEO had clearly issued a call, but the call was only inspirational. He had a vision, but it was not formulated in a manner that carried any personal significance for his people. He communicated what to believe, but he offered his people no direction on how to behave. Although this CEO made all the right noises about quality, customer focus and the like, *he did not provide his people with a clear set of factors that operationalized his vision.* The CEO's error was not that he made an impassioned plea for customer-driven quality, but that he failed to provide employees with a methodology for accomplishing it. He laid out lofty goals but offered no mechanism by which the brave new world would be achieved. Consequently, his people did not know what behavior was expected of them — individually or collectively.

In our experience, the CEO who launched his quality effort with the whirlwind, worldwide tour is representative of the many CEOs we have interviewed over the years. He proved very adept at communicating what was needed for success; he was less adept at defining for employees what they needed to do to achieve the goals and fulfill the vision. He made the two critical mistakes described in Chapter 3. First, his communication focused on changing attitudes rather than changing behavior. Second, he failed to provide a management system that everyone in the company could use to accomplish the vision of creating customer commitment. In short, he failed to move his quality initiative beyond the Rhetorical Component into the Activist Component.

The Five Whys and a Management Process

The problem faced by the CEO in the example is the same problem we posed at the beginning of this book. He didn't know whether the time, the money and the resources he and his company were investing in quality were paying off. The technique of the "Five Whys" provides a useful method for arriving at the root cause of the CEO's mistakes.[1] Using the Five Whys, one addresses a problem by asking a series of "why" questions until one arrives at the root cause of the problem. Once the root cause is uncovered, it can be addressed and solved.

"The key to our success has been the strength of the relationships we have with our customers and vendors. The Quality Review process has assisted us in enhancing these relationships while implementing a quality initiative that starts with the customer."

Tony Ibargüen,
Executive Vice President,
ENTEX Information Services

Problem:	**Companies today don't know if their quality investments are paying off.**
First Why:	Why don't corporations know if their quality investments are paying off?
Answer:	Because management has no way to measure quality in business terms.
Second Why:	Why doesn't management have a way to measure quality in business terms?
Answer:	Because quality is regarded as a parallel activity to the ongoing business practices of the company; it is thought to be only marginally related to business results.
Third Why:	Why is quality regarded as a parallel activity to ongoing business practices?
Answer:	Because the company is internally focused on improving what it does instead of being externally focused on what customers need and expect.
Fourth Why:	Why, if customer information is so vital, don't more organizations use customer requirements and expectations to drive their quality initiatives?
Answer:	Because organizations don't have a customer management system that allows them to use customer information the same way they use, say, financial information.
Fifth Why:	Why don't companies put a system in place that allows them to use customer information to improve the capability of their organizations?
Answer:	They don't know how.

At this point we have arrived at the problem's root cause — companies simply do not have the right tools to gather or use customer information, nor do they have a methodology for measuring and developing customer relationships. We have heralded the Quality Review process as a customer management system that addresses those issues. In the remainder of this chapter, we apply the principles of the Quality Review process to building both external and internal customer relationships.

First, we explore using the Quality Review process as a framework for communicating external customer requirements throughout the organization. Once employees understand external customer requirements, the Quality Review process methodology can then be applied to identify internal customer requirements. The latter part of this chapter looks at how the Quality Review process can be applied to help employees identify their internal customers; identify those customers' requirements and expectations; identify the means to meet those requirements and exceed those expectations; and finally, anticipate and meet their customers' needs.

"So how do you capture the customer? You capture him by engaging that customer in a dialogue ... [that] takes place in the language of service, support, information."

Industry Week
March 16, 1992

The Step from Theory to Practice

The Quality Review process addresses all of the issues raised by the Five Whys. It is a system that puts customer information on equal footing with financial and operational information.

Imagine a business without a financial system. That business simply produces products or delivers services, collects payment and writes checks to cover expenses, assuming that as long as the checks don't bounce, the business must be okay. Of course, that notion is ridiculous. But now imagine operating a business without any system for tracking information about the people who supply the revenue that keeps the financial system running — customers. As long as the revenue is coming in, customers must be satisfied, right?

Wrong. It is as ridiculous to run a business without customer information as it would be to run a business without a comprehensive financial system. Yet while corporations employ entire armies of accountants and financial analysts, few employ anyone dedicated more than incidentally to the acquisition and use of customer information. We have found many companies, Fortune 500 included, that could not even produce an accurate and current list of customers.

Yes, companies do employ people in the area of market research. Others spend six-figure sums to track customer satisfaction. But market research and aggregated customer satisfaction mea-

"In the 1980s, we saw a customer in every individual,' Jan Carlzon says. 'In the 1990s, we see an individual in every customer.' In two brief sentences, the charismatic president and CEO of Scandinavian Airlines (SAS) articulates his vision of world-class service quality in the next decade — a strategy that recognizes each customer as a unique human being with individual needs and expectations."

Forbes
September 23, 1991

sures are generally top-of-the-Ferris-wheel views that provide generic, impersonal perspectives on customers. They are the wrong tools for providing an organization with the intimate knowledge of individual customers required to modify the behavior of its employees so as to directly impact the customer relationship and drive business results.

The Quality Review process, on the other hand, focuses on changing behaviors. It is a mechanism by which a company can move up the Mission/Vision Hierarchy to the visionary objective of customer commitment. It is the means by which a company can actively manage customer relationships, treating them as valuable corporate assets. Over time, use of the Quality Review process provides a measurement and documentation system to help determine if your investment is paying off by linking quality initiatives with accepted business results such as revenue gain, profit improvement, market share increase, productivity gains, cost reductions, reduced cycle time and defect reductions.

Implementation of the Quality Review process is logical and straightforward (Figures 6.2, page 77, and 6.3, pages 78 and 79). The concept of the process is simple and provides real benefit by organizing customer contact into a structured process that everyone in a company can follow.

First, the Quality Review process helps define an organization's customer base by identifying the key customers in terms of their strategic importance (Step 1). Along with the organization's existing business measures (such as increases in revenue, profit, market share and others by which the organization determines its level of success), this information from Step 1 is used as a reference to determine key quality indicators that define the performance of the organization through the perception of its customers (Step 2).

The key differentiator of the Quality Review process lies in Steps 3 and 4. Rather than rely on a third party to survey customers on the quality indicators identified in Step 2, the organization uses the indicators as guides and trains employees to effectively conduct interviews with key customers. Following each interview, individual employees respond to immediate customer needs (Step 4b).

Using a company's line employees to interview customers makes the Quality Review process both a tool to gather and use customer information and a methodology for measuring and developing customer relationships. Employee-conducted interviews integrate quality into the everyday jobs of employees. They focus employees externally on the requirements and expectations of the company's customers. And, they instantaneously supply employees with action-yielding information about individual customer contacts.

In Steps 5, 6 and 7, the results of the interviews are processed and the report of the findings is used to make macro-level improvements. Those improvements are then evaluated in terms of the organization's key business objectives. (This evaluation may indicate the need to revise quality indicators or adjust the quality processes within the organization's standard operating procedures.) Then, results of the improvement effort are communicated internally to the organization and externally to customers (Step 8), and the process is repeated.

Figure 6.3 (pages 78 and 79) provides a detailed overview of the Quality Review process, describing the purpose and outcomes of each step. Quality Institute resources are available to facilitate each step, maximizing a company's investment in the process.

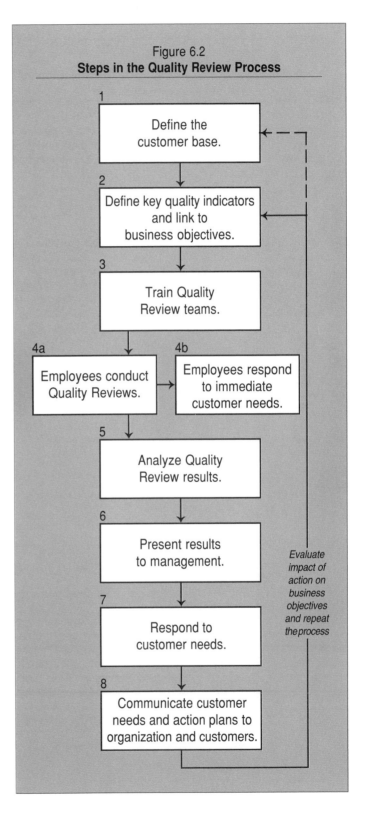

Figure 6.2
Steps in the Quality Review Process

1 Define the customer base.

2 Define key quality indicators and link to business objectives.

3 Train Quality Review teams.

4a Employees conduct Quality Reviews.

4b Employees respond to immediate customer needs.

5 Analyze Quality Review results.

6 Present results to management.

7 Respond to customer needs.

8 Communicate customer needs and action plans to organization and customers.

Evaluate impact of action on business objectives and repeat the process

Figure 6.3 • **Steps in the Quality Review Process (detail)**

STEP	QI AVAILABLE RESOURCES*	EXPECTED OUTCOMES
1. Define the Customer Base The organization identifies its customers by company and by key contacts within each company. All key customers should be interviewed.	**Customer Planning Document** The *Customer Planning Document* is a worksheet that provides an organized approach for identifying customers and the characteristics of the customer base.	**Customer Database** A systematically organized repository of customer information is formed, providing a clear picture of an organization's customers. This list is used to more effectively manage customer relationships.
2. Define Quality Indicators A cross section of the organization's employees identifies the factors of greatest importance to their customers' business success and links them to the key business objectives of their own organization.	**Quality Review Design Session** A design session is conducted with a select group of employees, management and (if appropriate) customers. During this session, a *Quality Indicator Identification* form facilitates the efficient design of the interview guide.	**Link to Business Objectives** A consensually obtained set of factors most important to customers is developed, providing a structured tool for tracking customers' perceptions of the organization's performance and for evaluating "quality processes" in terms of business objectives.
3. Train Interviewers Employees who are selected to conduct customer interviews are trained in the effective use of the structured interview guide.	**Training Workshop** A video-supported half-day workshop, *Conducting Quality Reviews,* is recommended for employees and for trainers.	**Trained Interviewer** Employees understand the rationale and techniques for interviewing customers, gathering complete customer information, reducing interviewing bias and enhancing reliability of results. Employees understand the applications of the process within their work group.
4. Conduct Quality Reviews Employees schedule appointments with customers and conduct face-to-face interviews. Employees are the direct recipients of their customers' feedback.	**Structured Interview Guides** The *Quality Review Interview Guide*, based on the quality indicators identified in Step 2, is used to gather information from customers about how well the organization is meeting customer requirements, the importance of those requirements, the rate of improvement, and the organization's competitiveness. Immediate feedback is provided to the interviewer in the *Contact Profile.*	**Relationship Management** A base of knowledge is formed about the organization's performance on those factors most important to its customers. Enhanced relationships are a result of employees' personal contact with customers. Immediate customer needs are addressed (Step 4b) and documented.

* Italicized terms indicate proprietary products available through Quality Institute.

Figure 6.3 • **Steps in the Quality Review Process (detail)**

STEP	QI AVAILABLE RESOURCES*	EXPECTED OUTCOMES
5. Analyze Review Results Completed interview guides are processed, a database is established and reports describing the results are produced.	**Database Management** A *Quality Information System* processes, tracks and trends information over time, provides comparison to a normative database, and produces an easy-to-read *Quality Review Report of Findings*. A *Customer Confidence Index™* is calculated to identify which customer relationships are at risk.	**Quality Review Database** A database of customer information is developed and used to prioritize actions for improvement, track progress toward achieving business objectives, and develop strategies for improving individual customer relationships.
6. Present Results to Management An executive summary of all results is presented to management. Priorities are identified and recommendations for actions are made. Managers become sponsors for specific strategic action items.	**Presentation and Content Analysis** An executive summary and content analysis is prepared. This concise presentation of quantitative and qualitative data highlights the customer data collected by employees and the actions taken to address immediate concerns.	**Management Sponsors Identified** A management focus on customer perceptions of the organization's performance is established. This focus provides the impetus for management to sponsor the strategic action planning related to customer priorities.
7. Respond to Customer Needs Management and selected customer action teams (including employees and others, e.g., customers, suppliers) develop and implement action plans to respond to needs expressed by customers. Actions are evaluated in terms of business objectives.	**Action-Planning Workshop** A second workshop, *Quality Improvement Planning,* is recommended. Training for employees and for trainers is available.	**Improvements Implemented** Specific short- and long-term action plans that address meeting customer requirements, exceeding expectations and anticipating needs are jointly developed by employees and management. Plans are developed for individual customers and across customer groups.
8. Communicate Plans The voice of the customer is disseminated throughout the organization and improvements are communicated to the customer base.	**Sample Communication Pieces** Examples of brochures, letters and other communication techniques used to alert other members of the organization to customer concerns and to communicate directly with customers are provided in the *Quality Improvement Planning* workshop. Implementing TALK[2] is the most effective way to communicate customer feedback internally.	**Closing the Loop** Everyone in the organization is aware of the impact of his or her actions on customers. Consequently, everyone in the organization is more focused on enhancing customer commitment. All customers are aware of the improvements made. Customer relationships are strengthened.

* Italicized terms indicate proprietary products available through Quality Institute

"Every action taken each day by each employee has an impact on the customer — either positively or negatively. If employees begin to place themselves in the customer's position, we can achieve our goal of total customer satisfaction."

Jim Weber
Manager of Development
Allen-Bradley Corporation

Differentiators of the Quality Review Process

Writing in his book *Kaizen*,[2] Masaaki Imai notes that there can be no such thing as a static constant. All states are destined to deteriorate once they have been established — and that includes the state of the relationships a company has with its customers and the relationships a given department in the company has with another. In other words, relationships with external and internal customers are in a constant state of flux. Consequent to Imai's principle, there must be a continuing effort for improvement to even maintain the status quo of the relationships. Without such efforts, decline is inevitable,[3] making the external customer vulnerable to defection and creating a state of disharmony and contention between internal customers and suppliers.

The repetitions of the Quality Review process are the small steps that lead to continuous improvement of an organization's relationships with its customers. They also lead to continuous improvement in the ability of an organization to define customer requirements and translate them into action plans. As employees become more adept at interviewing customers, they will also become more adept at anticipating and fulfilling customers' needs and expanding the business through increased sales of existing products and through exploitation of new opportunities uncovered through customer interviews.

Implementation of the eight steps of the Quality Review process represents a proactive effort on your part to obtain and disseminate customer information, as opposed to passively accepting a random flow of information into your company. The Quality Review process is differentiated from traditional customer research in five primary ways. These differentiators can be expressed as five advantages.

Customer information is disseminated directly to people who can act on it.

The Quality Review process is based on the premise that employees of the supplier (your company or your department) conduct interviews with its customers. Contrast this approach with hiring a third-party research firm to contact external customers or conduct an employee survey. The Quality Review approach has the advantage of customer information flowing directly to the people who need it — employees who are in a position to drive necessary changes. Further, the information is available without the time lag created by first summarizing the information to management and only then passing it down the chain to employees, as is done in traditional approaches.

Employees who can effect change are the first to know the requirements and expectations of customers, not the last. A corollary to that notion is that information employees hear from customers results in less resistance than information heard from management or a third-party research firm. Resistance occurs because when management "identifies" opportunities for improvement, it often sends the message (albeit unintentional) that employees are not trying. A lot of employee energy then goes into refuting that notion — energy that could be better spent improving customer relationships. On the other hand, hearing the same information directly from customers carries the ultimate credibility. Employees are motivated to take action immediately.

The often-ignored flip side of hearing negative feedback from customers is hearing positive feedback from customers. In terms of process management, positive feedback tells a company which processes are working. From an employee-morale perspective, reinforcement is enhanced when compliments come directly from customers — compliments that without a mechanism like the Quality Review process would probably never be paid. Numerous studies have validated what

What are we going to do about Hazel?

Just when the company president thought she had things going her way, the director of engineering had to mention Hazel. A long-time and valuable employee, Hazel had risen through the engineering ranks long before there were a lot of women in the field. She was good and she knew it, which tended to make her (to be ever so diplomatic) just a tad abrasive in her dealings with her co-workers and management — *and* in the rare instances when she unavoidably came in contact with external customers.

"Surely," said the director, "we aren't going to *voluntarily* put Hazel in front of customers?"

Citing employees with "less than exemplary" people skills is just one objection raised to the implementation of the Quality Review process. When an objection to the process is made based on the personality of a specific person, it is usually an instance of using a worst-case example to illustrate the more general fear that "we can't put *any* employee in front of customers." The executive who champions the Quality Review process is also likely to hear:

- Our employees don't have time to talk to customers. They're too busy doing their jobs.
- Customers are too busy to talk to our employees.
- We've just introduced a new quality initiative and we don't want to introduce a new philosophy/language/training. We don't want to shift gears now.
- We're already doing a customer satisfaction survey. Why do we need this?

Our experience finds that these objections are, in general, an indication of a company's internal priorities, not external realities. We will examine these objections and others in more detail in Chapter 11.

all of us know intuitively — people are more apt to complain about a negative experience than recognize a positive one. The direct Quality Review process question "What do we do well?" prompts positive recognition, which in the long run is as vital to an organization as learning about customers' negative experiences.

The information an organization receives directly from customers can be used to take action.

Regardless of a third-party researcher's interviewing skills, he or she can never know a business as well as a company's own employees know their business. When an organization's employees interview customers, their knowledge of the business allows them to probe for very specific details about customer requirements and expectations and the supplier's performance. The organization's employees know when to seek clarification, and they understand what questions need to be asked to prompt the customer to provide specific suggestions for improvement.

"The [Quality] Reviews are allowing us to uncover customer needs and issues that may not have been discussed previously.... We are gaining valuable insight into how to better service the needs of our customers."

Sheri Bojanowski, Manager
Cable & Wireless
Communications

The Quality Review process yields a high participation rate.

Third-party surveys use a number of techniques to ensure high and statistically valid response rates. The advantage of the Quality Review process is that the "response rate" is virtually 100 percent. Customers are always willing to talk with a supplier organization's employees because employees have the ability to impact the customer's business. If customers won't talk to a supplier's employee, the supplier has gained some valuable information — the customer has higher priorities than talking to the supplier's people, and the supplier probably has a relationship that is in trouble.

Customer receptivity to surveys, always tenuous, gets more so every day as the number and frequency of surveys increase. People are filling out more and more surveys and seeing fewer and fewer results — not only astronomers worry about "black holes." However, when a customer understands that a company employee — not a third party — is asking the questions, that customer is more likely to respond and to respond frankly. The customer has more confidence that the interview will result in action.

The organization can identify specific customer relationships that are in trouble.

Some of the most common results of third-party customer measurement are such statements as, "Overall, 15 percent of customers are dissatisfied with your customer service program." The obvious question is, "Which customers?" The Quality Review process provides a method for aggregating data for purposes of trending over time and across the customer base — the same kind of overall data provided by traditional methods. However, the real benefit of the Quality Review process, which comes about because an organization's employees are interviewing customers, is that information is immediately linked to the customer who provided it. The Quality Review process not only provides data for high-level strategic planning across the customer base, but also provides customer-by-customer information for tactical account management. If a customer relationship is in trouble, as indicated by the Customer Confidence Index™ score, the employee can take immediate steps to remedy the situation.

The Quality Review process measures customer confidence, not satisfaction.

"Confidence," defines Webster's, is "trust or faith, a trusting relationship." When one considers the issue carefully, it becomes obvious that confidence is the single most important factor in a customer's purchasing decision and subsequent loyalty. Without confidence in the supplier's ability to meet his or her needs, there certainly can be no relationship and probably no initial sale.

But what is the operational definition of "confidence"? Like "quality," "confidence" is an inherently vague term that has traditionally lacked an operational definition that makes it measurable. One reason for this is that confidence is not a uni-dimensional concept. It is multi-dimensional, resulting from the complex interactions of three key factors:

- How well the organization's present level of service or product *conforms* to customer requirements. (Most traditional customer measurements stop here. They ignore the following two crucial pieces of information.)

"During a Quality Review with a customer we had lost, we learned both about how they made the decision that led to losing the bid for renewal, and the advantages our competitors had. We were told precisely what we needed to do to win the account back. We have totally revised our approach with this account and now have been invited back to present to them. I am confident we will win this account back in the next 12 to 24 months."

Annette Wagener
Senior Director, Sales
HealthPartners

The Customer Confidence Index™:
A Measure of Conformance, Importance and Improvement

While mathematics makes customer confidence measurable and statistically accurate, common sense also validates the Customer Confidence Index as a valuable management tool. Use of the index has established the following commonsense interactions:

- Conformance has the greatest impact on customer confidence. If conformance is not high, confidence cannot be high.
- Greater conformance and greater improvement yield higher confidence.
- Greater importance increases the impact of conformance (positive or negative). As the importance of an attribute increases, low conformance is tolerated less.
- Indicators of little importance to customers cannot generate great confidence. Confidence cannot be calculated when importance is rated "not at all important."
- Improvement and confidence are positively correlated regardless of the conformance rating.

If performance improves, confidence increases. If performance declines, confidence decreases. With extreme ratings on conformance, however, the impact of improvement on confidence is marginal.

- Decline in improvement can damage confidence more than an increase in improvement can enhance confidence.
- When conformance is rated "sometimes conforms," the effects of improvement are magnified.
- The improvement rating "stayed the same" yields high confidence when conformance is high, and yields low confidence if conformance is moderate or low.

These interactions are reflected in a mathematically derived formula and are presented in terms of a numeric scale from 0 to 100.

- Whether the current level of service or product has *improved*.
- How *important* a particular aspect of the service or product is to the customer.

Unique to the Quality Review process, Quality Institute's Customer Confidence Index allows a company to quickly identify business opportunities on both an individual customer basis and across a market segment.[4] It also allows a company to quickly identify customer relationships that are in trouble. In both cases, the business results are increased revenue, profit and/or market share.

Quality Institute's Customer Confidence Index provides a mathematically valid way to calculate a customer's confidence in the supplier organization. The index provides an organization with the ability to trend the state of customer relationships, prioritize needed improvement and identify customer relationships that may be at risk.

Consider how the priority for improving a quality indicator changes using the three pieces of customer data incorporated into the Customer Confidence Index when compared to the single dimension of traditional customer satisfaction surveys. When only one piece of customer data is collected (such as conformance), false assumptions about the strength of the relationship are easily drawn. However, the prioritization provided by the Customer Confidence Index allows targeted action planning to proceed. The Customer Confidence Index helps an organization identify which customers are at risk, which are vulnerable to the competition and why. Once an at-risk account is identified, the company can *immediately* take steps to improve the relationship and retain the customer — and by doing so, reap the business benefits that are obvious from preventing a customer defection.

The Customer Confidence Index can be used to evaluate customer confidence on a single quality indicator or across a broad range of indicators, which provides an overall measurement of the customer's confidence. It can be applied to a single customer, or it can track quality indicators across an organization's entire customer base, magnifying the impact on business results.

What differentiates the Quality Review process from traditional models lies not in the basic elements of the model — customers, suppliers, employees — but in the interaction of those elements. Customer requirements do not magically flow into a company from external customers or into a department from internal customers, nor do they come neatly categorized. Nor does customer information automatically flow to the people who need it. That is the reality, the difference between theory and practice. What differentiates the Quality Review process is that it clearly delineates a system for actively managing the collection, analysis and dissemination of customer information. *The Quality Review process is the step from customer measurement theory to the practice of customer measurement.*

Using the Quality Review Process Internally

In Chapter 5 we noted that building employee commitment to the vision of creating committed customers is often missing from organizations' quality initiatives. Companies infected with the customer focus bug overlook the fact that customer commitment depends on employee commitment to customers. Employees are the crucial link between company and customer. Strong relationships between employees and customers depend on strong relationships between employees within the company. The Quality Review process used to build relationships with external customers can be easily adapted to build and manage internal customer/supplier relationships as well.

By now, the concept of "internal customer" should be very familiar. Quality gurus and consultants have effectively driven home the point that everyone in a company is both a supplier to the next process in line and a customer of the preceding process. But while most people intuitively understand the reason for managing external customer relationships — external customers have a choice — they are often less inclined to put effort into managing internal customer relationships. After all, *internal* customers are usually *captive* ones. They often cannot readily go elsewhere for the services or products provided them. So what is the motivation to spend a lot of resources on improving internal customer relationships?

Above all else, our research has found a strong correlation between internal cooperation and customer satisfaction. When the relationship between internal customers and suppliers is less than satisfactory, the external customer ultimately suffers. External customers will get the message loud and clear if the internal relationships in the organization are ineffective.

In light of this strong correlation, the following data takes on significant implications. Analysis of the employee data we have collected reveals only 23 percent of employees believe the cooperation between their work group or department and others is good or very good. On the other hand, 36 percent rate interdepartmental cooperation as fair, and 41 percent rate it as poor or very poor (Figure 6.4).

Employees have strong feelings about the reasons behind those numbers. Three in four employees (78 percent) say there is not enough communication between work groups and departments (Figure 6.5). Barely one-third (39 percent) rate the quality of interdepartmental communication as good. Only 29 percent of employees feel their internal suppliers understand their needs. Consequently, it is not surprising that only 24 percent rate the quality of products and services provided by internal suppliers as good (Figure 6.6, page 88).

If, as our data suggest, work groups recognize and understand that interdepartmental cooperation is a problem, and if they understand the consequences, then why don't they solve the problem? Once again, the Five Whys lead us to the conclusion that they simply don't know how.

We submit that the common objectives for internal customers and suppliers are the requirements, expectations and needs of the *external* customer. If the fundamental mission of an organization is meeting customer requirements and exceeding customer expectations, if the fundamental vision is creating committed customers (by anticipating needs), then the triad of customer requirements, expectations and needs is the only acceptable framework for internal customer/ supplier negotiation.

From Theory to Practice Again

Understanding external customer requirements, expectations, and needs is important only to the degree that understanding motivates behaviors. In other words, an *attitude* of confidence in the company's vision is not the same as *behaviors* that reflect that confidence. In a company, behaviors are driven by company processes and procedures. Processes and procedures are where quality theory translates into practice once again — this time from an internal perspective. And once again, the Quality Review process comes into play.

Using the Quality Review process to evaluate internal customer/ supplier relationships is simply an extension of the process as used to determine external customer requirements, expectations and needs. The consistency of methodology makes identification of

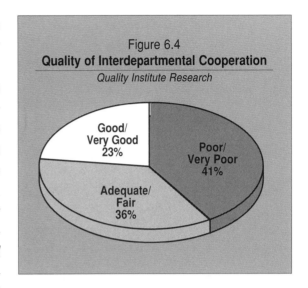

Figure 6.4
Quality of Interdepartmental Cooperation
Quality Institute Research

Good/ Very Good 23%
Poor/ Very Poor 41%
Adequate/ Fair 36%

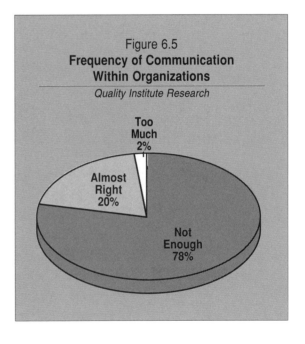

Figure 6.5
Frequency of Communication Within Organizations
Quality Institute Research

Too Much 2%
Almost Right 20%
Not Enough 78%

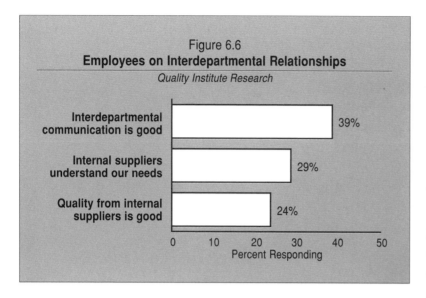

Figure 6.6
Employees on Interdepartmental Relationships
Quality Institute Research

internal customer/supplier relationships less arbitrary — internal relationships are based on external customer needs.

But who are the internal customers? As a rule of thumb, customer/supplier relationships work backwards from the external customer. Which internal organization is serving the external customer most directly? Chances are, that organization is the internal customer.[5]

For example, while it is true that staff organizations are customers for information provided by line organizations, the reason staff organizations exist is to *serve* line organizations. The true internal customer is the line organization.

Internal implementation of the Quality Review process follows the same eight steps as the external application (Figure 6.2). The differences between application of the Quality Review process with internal customers and external customers lie in content rather than methodology. In both cases, a design session is held to determine the quality indicators to be measured. Our experience indicates that in most cases internal quality indicators will be similar to, if not duplicates of, the quality indicators for external customers. However, enough cases of difference exist that congruence should not be assumed; quality indicators should be confirmed by the internal organization's employees in a design session. As is the case with external application of the Quality Review process, interviews are conducted by a cross section of the supplier department's people.

A systematic rollout is crucial when implementing the Quality Review process internally. Imagine thousands of mousetraps in a 10' × 10' square with a Ping-Pong ball poised on each. Toss a Ping-Pong ball into the middle of the square and you set off a chain reaction of popping Ping-Pong balls. Simply toss the Quality Review process into a company, and you could wind up with half your people interviewing the other half, paralyzing the organization. An internal application of the Quality Review process must never lose sight of the common objective of meeting *external* customer requirements and expectations. A systematic rollout of the process begins with that objective in mind.

As with the external implementation of the Quality Review process, the internal application is simple and displays good common sense. A structured form of communication between employees of an organization is intuitively logical. When implemented in an orderly, systematic manner, the process clearly directs the behavior of every employee toward creating customer commitment.

"The lesson is clear. Our reputation is based on what we have done. But to continue to be perceived well, we have to focus on what we need to do better — not so that it paralyzes us, but so that it motivates us to improve."

L.D. DeSimone
President and CEO
3M Company

W e began this chapter with a CEO who had a problem. He had just traveled around the world promoting his vision of quality, but he failed to operationalize the vision. We might very well end the chapter with a problem faced by another CEO, who chose to operationalize her vision using the Quality Review process. Convinced of the importance of customer information and convinced that the Quality Review process provides a method for effectively managing customer relationships throughout her organization, this company president found herself in a not uncommon position for a quality visionary — facing resistance to a new approach (see "What are we going to do about Hazel?" on page 81).

Returning to an idea introduced in Chapter 4, the Quality Review process as a vision of a new way of managing customer relationships is both an extension of and a departure from the status quo. While it is based on sound research principles, the notion of putting line employees in direct contact with customers is a clear departure from the tradition of third-party surveys. Some people are uncomfortable with that change. They envision the Quality Review process as a threat simply because it represents a change from the familiar — and change breeds objection.

This discussion does not invalidate the objections raised in the example. The point here is that for the most part, the Hazels of the world are made, not born. Their attitude toward customers is formed by a company's ongoing business practices, policies and priorities as demonstrated by daily business decisions. Those items define for employees the company's perceived mission, and they are reflected in employee attitudes and behaviors toward customers.

Objections to allowing employees to interview customers indicate that the "real" mission of the company is something other than meeting customer requirements and exceeding customer expectations. Customers are not regarded as a priority of people's everyday jobs. Consequently, customer contact is not managed. The result is ad hoc behavior patterns in dealing with customers. Employees interact with customers from their perspectives (internal focus), not from the customer's perspective (external focus), because their own perspectives are the only ones they know. The existence of the objection, "We can't put Hazel (or Joe, or Jane) in front of customers," is a concrete indication of *the need for a process that puts people (even Hazel!) in touch with customer requirements and expectations* — the Quality Review process.

[1]   The "Five Whys" methodology was developed by Taiichi Ohno of Toyota Motors.

[2]   Masaaki Imai. *KAIZEN: The Key to Japan's Competitive Success.* Copyright 1986 The KAIZEN Institute, Ltd. "KAIZEN" means gradual, unending improvement, doing "little things" better; setting, and achieving, ever-higher standards.

[3]   Imai. *Ibid.*

[4]   The Customer Confidence Index™ was developed by a team of professionals with multi-disciplinary backgrounds that include psychology, sociology, communications, marketing and mathematics. The index was developed after a comprehensive review of the research on customer satisfaction, retention and commitment, and the completion of a number of primary research efforts (focus groups, customer surveys, in-depth interviews and empirical pilot studies).

The Customer Confidence Index is grounded in several classic theories, including motivational theory, exchange theory, decision theory, expectancy theory and cognitive dissonance theory. These approaches to human behavior established the basis for the components and statistical equations used to calculate the index.

Expert panels were used to establish a Customer Confidence Index for every possible combination of the interaction between conformance, improvement, and importance. The statistical formula used to calculate the index was correlated with the findings from the panels. Stepwise multiple regressions were run against the panel data. Coefficients were rounded. The resulting predictions were again rounded to the nearest integer and raised or lowered to the scale boundaries when necessary. Models that could not be fully justified conceptually, were insufficiently sparse, or that were sensitive to rounding were eliminated. Residuals and squared residuals were examined for patterns and clusters. The correlation between the formula and the expert panels' scores is .995. A variation of the formula, used when conformance is rated in the mid-range, takes into account a magnified impact of improvement. Correlation of this second formula with the expert panels' scores is .996.

[5]   For the purposes of the Quality Review process, reporting relationships are not defined as internal customer/supplier relationships. However, Quality Review process principles do apply to reporting relationships and this application is discussed in Chapter 9.

"TALK² provides the process that gets managers and employees talking with each other and moving in the same direction."

7

TALK²:
Tapping Employee Expertise

WHAT WOULD YOU CALL THEM? TWO JACKASSES ARE TIED TOGETHER by a 20-foot rope. On each side of the corral where they are penned is a bale of hay. The two animals keep straining and tugging at each other, each trying to get the nearest bale. This struggle goes on until finally both of the animals collapse from hunger and die of starvation. If either jackass had been willing to cooperate, both could have eaten their fill from both bales — first from one and then the other.

That story was first told to us by a colleague who used it to describe the relationship between managers and employees at his company. "Both management and employees want to meet customers' needs," he noted, "but the two groups aren't working together. They're pulling in different directions."

We do not relate this story to add to the list of names employees have for management or by which managers disparage the work force. (Far more nimble minds than ours are at work on this effort

"The ability to manage work forces and create a workplace that empowers people and continually taps human potential are the challenges before us. The degree to which people choose to exert their best effort will determine the success of companies in a competitive global economy."

Frederick W. Smith,
CEO
Federal Express

every day, in just about every organization.) And while one may smile or laugh at stories about jackasses, it is likely a tense smile or a nervous laugh. We relate the story because we know from our research that savvy managers and perceptive employees recognize that the humor of the jackass story is but once removed from the significant damage poor management/employee relationships can have on an organization's ability to build relationships with its external customers.

As we saw in Chapter 5, relationship building and employee participation in the vision of the company are *not* two separate, serial issues. An employee relationship-building program managed separately from programs designed to tap into employee expertise is doomed. Such programs fail to recognize that commitment results *from* participation, that they go hand-in-hand.

The Quality Review process as discussed in the previous chapter is an example of using employee participation to build customer commitment. By having employees conduct structured interviews with customers, employees become intimately connected with the requirements, expectations, and needs of external customers. However, while connection with the customer is a crucial element in creating committed employees, it is not the only element.

Once they understand customer requirements, employees are anxious to put that new knowledge to work. Ideas start to flow. On one level, the question is how these ideas get uncovered and implemented. On another level, the question is how the *process* of uncovering and implementing ideas becomes integrated into the working culture of the organization. In this chapter, we'll look at TALK[2], a process for *tapping the expertise of the organization for ideas that contribute to business results.*

Why TALK[2]?

Realizing that "employee empowerment" is more than a buzz-word, managers have tried many ways to increase employee commitment to the organizational visions and missions. But despite all the time, money and resources spent on the effort, most organizations have failed to achieve the level of employee commitment required to gain and maintain a competitive edge in today's business environment.

Quality Institute research shows that one major reason employee commitment to the vision of committed customers falls short is a perceived lack of organizational communication upward, downward, and across the organization, both formal and informal. Specifically, when employees are asked to comment on basic communication issues, few respond favorably (Figure 7.1).

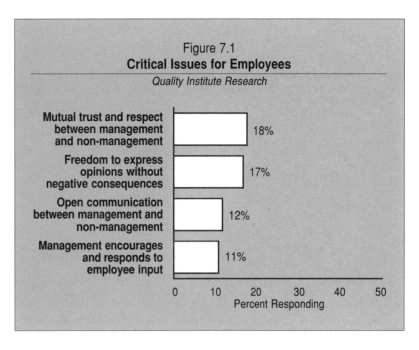

Figure 7.1
Critical Issues for Employees
Quality Institute Research

The perception of employees is that most organizational attempts to improve communication have failed. Yet, organizations seem to be spending more time, money, and resources on communication than ever before.

Why isn't this investment paying off? The answer is not necessarily a lack of communication, but rather a lack of the right *kind* — direct, one-on-one dialogue with management. Dialogue with a purpose. Dialogue that relays customer concerns to employees. Dialogue intended to tap employee expertise and solicit employee ideas for better ways to serve the external customer. Dialogue that encourages employees to take personal responsibility for making improvements that create committed customers.

Note that we are talking about *one-on-one* communication. Most organizations are adept at establishing top-down communication. Some have mastered the technique of "Ask the CEO" programs where employees submit written questions to executives, but few have been able to integrate regular employee/manager discussions into their day-to-day work activities.

Why is that?

Once again, we can apply the logic of the Five Whys and conclude that organizations' difficulty in establishing one-on-one communication between managers and employees stems not from a lack of will, but rather from the lack of a system. Personal one-on-one dialogue is not easy, and without a process to follow, most people will look for more comfortable ways to communicate — even if those methods prove less effective.

What Is TALK²?

TALK² is an innovative and simple process that helps an organization achieve its objectives by tapping the wisdom of its employees while improving management/employee relations. TALK² encourages openness between management and employees, enhances employee trust and confidence in the vision of the organization, reinforces commitment to common goals, and helps managers better understand and act upon employee ideas for improving organizational performance.

Very simply, TALK² is managers talking with their employees, one on one, using a structured interview guide to facilitate a discussion about making improvements in the effectiveness of the organization. At first blush, TALK² may seem very much like what is already happening in your organization. Indeed, some excellent managers intuitively use the techniques formalized in TALK² every day. The point is, most managers do not.

TALK² is a tool. As with any tool, it is effective only when used for its intended purposes — first, to focus employees on the needs of the organization's customers and second, to discuss employees's ideas for meeting customer requirements more effectively. This objective should not be confused by making TALK² into something it is not:

"The new economics of service requires innovative measurement techniques. These techniques calibrate the impact of employee satisfaction, loyalty, and productivity on the value of products and services delivered so that managers can build customer satisfaction and loyalty and assess the corresponding impact on profitability and growth."

J.L. Heskett, et al.
Harvard Business Review
March-April, 1994

- TALK² is not an individual performance review process; it is an organizational review process.

- TALK² is not a "rate your boss" discussion; it is a "rate the effectiveness of the organization" discussion.

- TALK² is not an anonymous employee survey; it is intended to be an open and interactive discussion.

Like the Quality Review process, TALK² goes beyond just gathering information. It sends a message of change throughout the organization. TALK² is the process that gets managers and employees talking with each other and moving in the same direction.

TALK² encourages dialogue at every level and across all functions of the organization. It makes use of existing hierarchies, releasing the energy and creativity of the organization as information "cascades up" the organization.

The TALK² Process

Figure 7.2 shows the eight steps in the TALK² process. In form and function, the process is very similar to the Quality Review process (Figure 6.2, page 77).

In Step 1, people representing the organization's broad range of job functions and levels come together to design the interview guide that will be used by

In Praise of Hierarchy

At first glance, hierarchy may seem difficult to praise. Bureaucracy is a dirty word even among bureaucrats, and in business there is a widespread view that managerial hierarchy kills initiative, crushes creativity, and has therefore seen its day. Yet 35 years of research have convinced me that managerial hierarchy is the most efficient, the hardiest, and in fact the most natural structure ever devised for large organizations. Properly structured, hierarchy can release energy and creativity, rationalize productivity, and actually improve morale. Moreover, I think most managers know this intuitively and have only lacked a workable structure and a decent intellectual justification for what they have always know could work and work well.

Elliott Jaques
Harvard Business Review
January-February, 1990

managers when conducting TALK² discussions with their direct-report employees. The interview guide summarizes the Quality Review information that has been collected so that each employee will be familiar with current customer feedback. It then seeks employee input on several topics directly related to how the organization might improve these areas. Managers solicit employees' ideas on specific issues such as:

- How to improve relationships with customers.
- How to improve the quality of the company's products and services.
- What barriers exist to improving performance.
- How well the company is meeting its goals.
- How effectively the organization communicates internally.
- How effectively the organization encourages the development of its people.

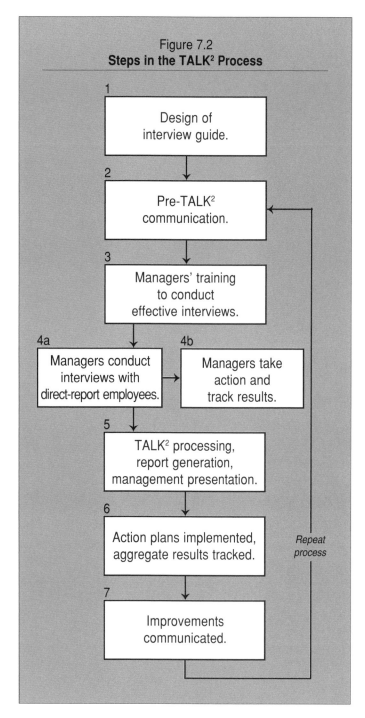

Figure 7.2
Steps in the TALK² Process

1
Design of
interview guide.

2
Pre-TALK²
communication.

3
Managers' training
to conduct
effective interviews.

4a
Managers conduct
interviews with
direct-report employees.

4b
Managers take
action and
track results.

5
TALK² processing,
report generation,
management presentation.

6
Action plans implemented,
aggregate results tracked.

7
Improvements
communicated.

*Repeat
process*

Step 2 of the process is informing the organization about the TALK² process *before* any interviews take place. This is not a trivial exercise. As mentioned before, TALK² is an innovative approach to employee/management communication. It represents change (radical change for some) and until the process is fully understood, it may be met with some trepidation, even resistance. Therefore, it is important that the organization clearly communicate the objectives and nature of TALK² (what it is and what it is not) to the organization before proceeding to the next steps.

Steps 3, 4, and 5 require direct management involvement and commitment. In Step 3, managers and supervisors learn to conduct structured interviews. The objective of this training is to provide both an understanding of the TALK² process and an appreciation of the concerns employees may have when participating in a TALK² interview.

Step 4 is the interview process itself. Interviews start with first-line supervisors; employees are interviewed by their direct management. In turn, these first-line supervisors are interviewed by *their* managers. During the interview process, managers will likely learn about improvements they can make immediately. It is important that they do so and publicize the fact (Step 4b). The credibility of the TALK² process hinges on employees seeing tangible results — that is, changes in the workplace — following the interviews.

Step 5 is where work group-level data is rolled up and communicated up through the organization. Although many improvements can be made immediately at the work group level, some potential opportunities will involve more than a single work group or require significant corporate resources to correct. Through managers' interviews with *their* management, these longer-term issues are raised to levels where action can be taken (Step 6).

Step 7 introduces the ongoing communication efforts, primarily tracking improvement and reporting results. Remember that success breeds success. As concrete business results occur that can be directly linked to the TALK² process, more-tentative members of the organization will be motivated to actively participate in the process.

TALK² and Business Results

Armed with information gained through the TALK² interview cycle, companies can, and have, achieved some remarkable results. Using the TALK² process, companies have been able to increase the efficiency of their internal processes and procedures and uncover new ways of delighting their customers. They have uncovered easily remedied sources of employee dissatisfaction that, once corrected, increase employee confidence in the company's ability to create committed customers. But perhaps the most significant result of the TALK² process is that it is instrumental in integrating customer feedback into the day-to-day activities of an organization.

"I have been able to sit with each employee uninterrupted for 30 to 60 minutes. Each employee expressed concerns, frustrations and opinions in a way that normal day-to-day operations usually do not allow."

Supervisor,
after conducting
TALK² interviews

A prime example of this integration is this manager's comment about TALK²:

"Even though we work in the same environment, I have a better understanding [after participating in TALK²] of each employee and how their perceptions can differ so vastly. We see things through our own eyes, so therefore we formulate our own opinions and ideas as to how things are going. So often we are too quick to judge, basing our judgment on how things are at another company or how, ideally, we would like for things to be. Past experiences also play a major role in how each individual views situations. This became very obvious as a result of these interviews. I have also recognized areas that I can focus on for improvement that were not even considered important to me prior to these interviews — but that was my perception."

"Since people actually make things happen, they must be at the heart of any effort to change and improve the ways business is conducted."

Lamb and Publow
Quality Progress
November, 1992

I magine each supervisor and manager, on every level of a company, maintaining a schedule for gathering information from employees. Imagine the process "cascading up" the organization — information that could be acted upon to increase the competitiveness of the organization as a whole. Imagine managers talking with employees, clarifying issues, gathering ideas for change, and immediately acting upon those issues in their control and communicating company-wide issues to their managers. Imagine managers and supervisors continuously working at developing their relationships with their employees. Imagine employees taking personal responsibility to meet customers' requirements and exceed their expectations.

Innovative? Revolutionary? We think so. Will there be objections to the idea? Certainly. Most often we hear "Employees won't feel comfortable being honest if they are talking directly with their managers." In some instances this may be true. And the extent to which it *is* true is precisely why change is needed — desperately, in some cases.

Is it realistic to assume that relationships can be improved and employees motivated to contribute to the organization in any *other* way than by management and employees talking directly with each other and sharing ideas? It may feel a little uncomfortable at first, but the impact of achieving the corporate vision of creating committed customers is clear.

"The current business environment is a ripe opportunity to move beyond the focus on managing suppliers to building relation- ships with suppliers — getting them away from the sidelines, off the bench, and into the game."

8

The Supplier Review Process

"**Y**OU CAN'T MAKE A SILK PURSE OUT OF A SOW'S EAR" IS AN APHORISM of commonsense wisdom. Yet it is surprising how many companies try to create quality products from porcine procurement — companies dedicated to meeting customers' requirements who nonetheless accept decidedly inferior products from their own suppliers. At the other extreme are the companies that simply pass on their own internal problems to their suppliers. To compensate for the high variability in their own processes, these companies apply rigorous requirements on their suppliers and then establish elaborate vendor certification programs based on the ability to meet those standards.

In Chapter 5 we emphasized the point that customer relationships are true corporate assets. In today's highly competitive environment, the same holds true for a company's supplier relationships. Building and managing a company's relationships with its suppliers is vital to the organization. Quality suppliers are in demand. In many industries, the best suppliers can pick and choose with whom they will do business. And a company can improve its internal processes only so far before the only practical source of quality improvement lies in

Customer Influence on Supplier Quality

Minnesota companies, which have hauled in the Oscar of the business world [Malcolm Baldrige National Quality Award] two years running, have a good crack at landing more of them.

The reason: Supplier chains knit many of the state's businesses together. Efforts to improve quality at one company can spread through the supply channels like water seeping through a parched field.

Quality junkies throughout the state jumped for joy when Eden Prairie-based Zytec Corp. won the coveted [1991] Baldrige award for excellence in managing quality. [In 1990,] IBM-Rochester won.

Zytec CEO Ron Schmidt and his troops snared the Baldrige all by themselves. In driving toward the award, they showed the moxie of a Norman Schwarzkopf knocking out Saddam Hussein. Yet one fact can't be overlooked: Zytec happens to be one of IBM-Rochester's key suppliers.

The Eden Prairie company has produced power supply units for IBM-Rochester since 1986. After IBM decided to farm this work out, engineers from both companies worked together closely to design the units. Zytec people went to Rochester; IBMers went to Eden Prairie and to Zytec's factory in Redwood Falls.

"I think Zytec got a lot of encouragement and help from IBM-Rochester," says Jim Buckman, president of the Minnesota Council for Quality. "Companies in the supplier networks are constantly raising the bar. Once a company attains one level of accomplishment, other companies encourage it to reach a higher level. I think a lot of that is happening in Minnesota now."

St. Paul Pioneer Press
October 14, 1991

increased improvements in the quality received from the company's suppliers.

Wait. What's wrong with this picture? Haven't we spent the better part of seven chapters telling you that as a supplier your company has an obligation to develop its relationships with its customers? Well, you say, what about my suppliers? Don't they owe me the same effort?

Sure they do. But taking refuge in that belief is about as useful as determining who had the right-of-way *after* an intersection collision. Responsibility may get assigned, but the damage is already done. The point is, if your suppliers aren't taking the initiative to manage their relationships with your organization, or they are doing it through traditional third-party satisfaction surveys, and you're not seeing any tangible behavior changes on their part, then logic dictates that you must take the initiative and develop a supplier relationship strategy.

A Question of Relationships

Like quality assurance, supply-line management has joined the pantheon of hot, newly discovered business functions. Organizations are spending considerable time, money and resources developing sophisticated supplier certification systems. They are tracking a variety of supplier conformance measures. They are even requiring that their suppliers apply for, or put in place plans to apply for, the Malcolm Baldrige National Quality Award and/or ISO 9000 Series certification.

All of this activity is not without positive effects. However, our database of employee survey results shows that only 27 percent of employees believe open, two-way communication exists between their companies and their companies' suppliers.

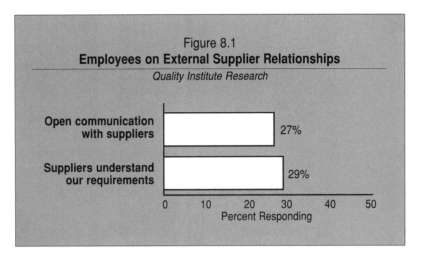

Figure 8.1
Employees on External Supplier Relationships
Quality Institute Research

Only 29 percent believe that their companies' external suppliers have a good understanding of the quality standards required of them (Figure 8.1).

Why the apparent discrepancy? It is a question of relationships. Supplier conformance to requirements, like customer satisfaction, is only one sign of the health of the relationship between an organization and its suppliers. As with customers, the strongest relationships with suppliers are based on five observable behaviors (as discussed in Chapter 5):

- Interactions between individuals are frequent and occur over a long period of time.
- Individuals are open to each other and are willing to disclose intimate information.
- Individuals develop effective communication systems and are willing to provide positive and negative feedback about each other's performance.
- Individuals work toward common goals.
- Individuals perceive that their relationship is unique and irreplaceable. They feel their individual interests are best served by maintaining a close partnership.

These are not the commonly stated goals of a supplier management effort. Typically, supplier relationships are managed by a procurement group — a single point of contact within the organization. As a result, supplier conformance is measured along a very narrow band of concerns. Larger impacts on the total organization are neglected. Interactions with individuals outside procurement are infrequent. Consequently, key pieces of information may not get exchanged. Feedback is usually based on who wields the most power — positive feedback ("politically" justified) flows from weaker to stronger, and negative feedback (as bartering leverage) flows from stronger to weaker. Supplier and customer are both internally focused on their own goals, and as our data implies, employees of the customer organization do not perceive any added value in their company's relationship with its suppliers beyond the products or service supplied.

The Quality Review Process and Managing Supplier Relationships

Just as with customers, the current business environment is a ripe opportunity to move beyond the focus on managing suppliers to building relationships with suppliers — getting them away from the sidelines, off the bench and into the game. A supplier version of the Quality Review process provides an opportunity for organizations to gain competitive advantage.

"In a boundaryless company, suppliers aren't 'outsiders.' They are drawn closer and become trusted partners in the total business process."

General Electric
Annual Report, 1991

As illustrated in the flow chart in Figure 8.2, the supplier review process addresses suppliers in the same manner as the Quality Review process addresses customers, involving as many members of the organization with suppliers as is relevant. Supplier performance is reviewed by a group of employees who are affected by the supplier's performance. Beyond the procurement department, this "Supplier Task Force" might include individuals from shipping and receiving who handle incoming product from suppliers, the accounts payable department that reconciles the supplier's invoices and the end users of the supplier's product within the organization. In some cases, it might also include people from engineering functions who incorporate the supplier's product(s) into their designs.

With guidance from the procurement organization, the Supplier Task Force compiles a list of all external suppliers, categorizes them based on their criticality to the organization's success, and selects those suppliers to be reviewed (Step 1). To achieve consistent results, the company's key quality indicators — the factors of greatest importance to the success of the enterprise — are identified. These indicators become the basis for the supplier Quality Review form that will be used to evaluate the supplier's performance (Step 2).

Employee teams are formed, made up of individuals who have direct experience with the supplier and/or the supplier's products and services (Step 3). The charter for each supplier review team is:

- Evaluate the supplier's conformance to requirements.
- Evaluate the improvement the supplier has made since the last review (approximately six months).
- Make specific improvement suggestions.
- Recognize their supplier's excellent performance.
- Rate each quality indicator in terms of its importance to the company.

Team members are educated on the charter and trained in the use of the supplier Quality Review form. Team members meet and complete the review (Step 4). Results may be formulated and presented to suppliers in one of two ways. One option is to present each team member's evaluation as a discrete observation (Step 5a), communicating back to the supplier that there may be differing perceptions of its performance within the customer organization. This is a good option when the supplier's products or services impact several different customer groups or functions. A second option,

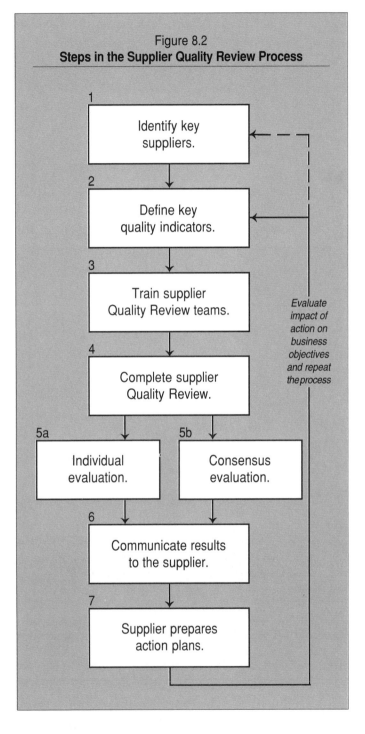

Figure 8.2
Steps in the Supplier Quality Review Process

"[Supplier] partnering promotes operations excellence through improvements in communication, quality, delivery, service performance, and total cost of ownership."

Sam Goldstein
former Director of
Vendor Relations
Rockwell International

applicable when the supplier's products or services impact a relatively focused group of customer functions (Step 5b), is providing the supplier with a single performance rating for each criterion.

Once the supplier review team has completed the supplier Quality Review form and determined how the data will be presented to the supplier, members of the team present the data to representatives of the supplier organization (Step 6). The ratings should be presented in terms of the costs to the customer organization caused by poor performance, or the savings realized when the supplier meets requirements and exceeds expectations. The nature of the review itself should be one of open exchange where dialogue is encouraged. The final item on the agenda is the customer's request of the supplier to prepare an action plan to respond to review issues (Step 7).

Integration of the Supplier Review Process

Supplier reviews should be conducted at regularly scheduled intervals — approximately every six months. As the participants become more familiar with the process, a gradual change occurs. Although frank feedback on performance remains a significant part of supplier reviews, these sessions gradually move away from strictly a performance appraisal of the supplier to an information-sharing session. The reviews move beyond a focus merely on requirements and expectations to a focus on future needs of the customer and future capabilities of the supplier. Review sessions facilitate the formation of mutual goals. The supplier/customer relationship will develop a value of its own, beyond the singular product/service focus.

And how will you know if this quality activity is paying off? Because of the process's one-on-one focus on individual suppliers, improvements made by suppliers as a result of the reviews can be linked directly to their impact on your business. This is a key quantitative measure of the effectiveness of the effort. However, there is a less tangible, but perhaps more significant indicator.

Recall the question raised early in this chapter — "Doesn't my supplier have an obligation to develop its relationship with my organization?" The answer is still "yes," and it defines the integrative goal of the supplier Quality Review process. The objective of the supplier review process *is its own elimination in favor of the supplier implementing its own customer Quality Review process with you.* The ultimate benefit of the supplier review process is motivating your suppliers to think in terms of managing their relationship with your organization — responding to your requirements and expectations, and anticipating your needs.

Clearly, the supplier Quality Review process is a departure from traditional thinking about supplier management — a paradigm shift, if you will, in thinking about supplier management. In Chapter 11, the issues of paradigm shifting and traditional research versus information gathered through the Quality Review process will be discussed in more depth. However, in summary of this chapter, two points need to be made about the supplier Quality Review process.

First, the supplier process does not intend to replace the existing supplier certification programs that may already be in place in your organization. Indeed, the criteria of such programs will (with little modification) likely become the key quality indicators used in the supplier Quality Review form. The supplier review process enhances, not replaces, existing supplier management initiatives.

Traditional supplier management efforts focus on measurement and numbers that reflect the degree to which the supplier is meeting your organization's requirements — the lowest levels of the Mission/Vision hierarchy, presented in Chapter 2. For your organization to achieve maximum leverage from a supplier, the supplier must elevate its vision and strive not just to satisfy your organization, but to turn your organization into a committed customer. The only way to do this is by creating value for your

---

Supplier Relationships

Traditionally, purchasing has dealt with suppliers at arms' length. Relationships have been played against each other in hopes of attaining lower prices. Consequently, suppliers have been unwilling to invest in the relationship and to improve products and services for fear of losing their investment along with the business. Even if suppliers were willing to invest in the relationship, customers have been reluctant to share ideas, information, technology and details of process for fear of it being used against them, making it difficult (if not impossible) for suppliers to assist their customers and make improvements. In this sense, relationships have been closed.

... Communication must be open and direct because purchasing can no longer funnel all communication.

... The role of purchasing becomes one of managing the relationship.

S. Ray
Total Quality, 1990

organization that goes beyond its products and services. Traditional supplier management does not engender that kind of thinking.

A second point that distinguishes the supplier Quality Review process from traditional supplier management is its integrative emphasis. The ultimate form of recognition in traditional supplier management is when a supplier achieves the highest level of certification a customer organization *bestows* upon its suppliers. A worthy accomplishment, but note how the event itself clearly defines *two separate entities* — the giver and the givee. The motivation for the supplier is to achieve certification, essentially an internal focus.

How different that goal is from the goal of the supplier who has integrated the Quality Review process into its way of doing business. The latter's focus is external on the goals of its customer — and ultimately that is the behavior you want exhibited by your suppliers. Your organization's benefit from conducting supplier reviews is motivating the supplier to conform to your requirements so you can conform to those of your customers. The supplier Quality Review process, through evolution into your supplier's own Quality Review process, lets your supplier see your customers as its own. That, we submit, is when your supplier management efforts truly begin to pay off.

"For quality to make a lasting impact, every department [in the company] has to be focused on customer commitment — not just satisfaction."

9

Does Quality Pay?
Look for Business Results

I T WAS HIS LAST SLIDE AND HIS BEST. WITH A TWITCH OF HIS THUMB ON the remote switch, the vice president of quality assurance splashed a kaleidoscope of color on the screen at the front of the room. Visually, it was a truly impressive slide.

"This," said the quality executive with a flourish, "is the culmination of our second year in quality. We have 50 teams meeting regularly, we have presented 10 "Big Q" awards to employees who submitted the best quality suggestions, and we have trained all employees in quality awareness and principles of Total Quality Management. This slide also shows a complete mapping of all of the quality processes that each plant is putting in place to address work process issues. In summary, I am pleased to announce that we are optimistic that we may begin to see some real benefits from this program in the upcoming year and beyond. Most of our plants are well along the way to establishing quality processes — with one notable exception, as you can see here."

The division president squinted at the slide. He turned toward the general manager of the division's most successful plant. "Jim," he said. "why is your group the only one that isn't meeting its quality process objectives?"

"Well," said Jim rather slowly, "at our plant, we really don't talk all that much about quality processes."

Every head in the room jerked up. Eyes bounced from Jim to the president, then back to Jim. It was no doubt only because Jim's plant made the largest contribution to corporate profit that the president hadn't gone ballistic on him. But it was only a matter of time.

"Let me understand you, Jim," the president said in carefully measured tones. "We've spent hundreds of thousands of dollars and untold employee time creating quality awareness in this company. Every other plant in this company is meeting its MBO[1] objective of implementing quality processes, and you're telling me that your organization isn't doing anything?"

"That's not what I said." Jim seemed surprisingly unruffled. "What I said was, we don't talk that much about quality processes. We're usually talking about our *business* objectives. For us, quality and business objectives are indivisibly linked. If you'd like, I have some transparencies that will show you how we're addressing our quality issues." He was already out of his seat and moving to the front of the room when the president nodded his head.

> *"The need to link quality initiatives to bottom-line results is no longer up for election."*
>
> Frank Miller, CEO
> Darling International

Quality Processes Equal Business Processes

Always a maverick within management circles, Jim was somewhat of a folk hero in the company. Five years earlier, he'd been given the unenviable task of turning around one of the company's poorest performing plants. Soon after his arrival, the plant made a shipment of such poor quality to a major customer that the company lost the account. That single event could have sounded the plant's death knell, but Jim used it as an example to his people. "There are two lessons we must learn from this experience," he told them. "We must get close to our customers, and we must use the information provided to us by our customers to improve the quality of our products." Jim's first transparency summarized the lessons of that early disaster.

"But we didn't stop there," he told the gathering of executives. "We went on to set up formal mechanisms to keep the people in our organization in front of our customers." He replaced his transparency with one showing a flow chart of the management system used for contacting customers. "We started out by asking customers how we were doing in meeting their requirements and expectations, but we quickly expanded from just asking them about their level of satisfaction to determining their level of confidence in our organization. We started asking for the kind of information we needed in order to anticipate their future needs. The more we listened, the more actions we took, the more customers told us and the greater their confidence in us became."

"But you must have changed some processes to put better quality into your products," interrupted the quality assurance vice president. "I know some of your people were in our SPC training. And you sent some people to our QFD training as well."

"Don't misunderstand," Jim responded. "Our plant supports and is thankful for a lot of the initiatives of your quality assurance group, Ed. And we fully support the company's efforts to become more customer focused. It's just that back when we were having quality problems, we did a lot of looking at other companies, trying to find a quick way out of our situation. What we found were companies that, in their enthusiasm to be 'customer focused' and 'market driven,' were caught up in a frenzy of creating new and elaborate processes. They formulated new objectives and overlaid them on existing objectives. Their employees were being 'MBO'd' beyond their capacity to achieve, with no central focus, and often with conflicting objectives that put them in a lose/lose situation right from the start.

"We looked at all the confusion that approach was causing — nobody in those companies really knew if their quality initiatives were providing any payoff — and arrived at a conclusion that seems like common sense now, but was anything but obvious back in our chaotic days. We looked at our operation and realized that we already were a functioning business. Granted, we weren't a profitable business, but we had some basic management systems in place. If we were going to *achieve* quality, we had to *integrate* quality and customer information into those existing systems. The lesson was clear: For quality to have a lasting impact, every department has to be focused on customer commitment — not just satisfaction — and customer commitment has to be integrated into the everyday, ongoing practices of the company's business.

"It quickly became obvious that we weren't going to achieve quality integration by introducing all kinds of new processes. New processes would only focus our people internally on learning new systems rather than externally on the requirements, expectations and needs of our customers. Quality has to contribute to the bottom line — not just by cutting waste and increasing productivity, but in a way that has impact on the customer and contributes to the organization's ability to grow the business.

"Focusing on customer relationships has forced us to implement initiatives, quality or otherwise, that positively impact our customers, initiatives that enhance the relationships we have with our customers, initiatives that reinforce our customers' commitment to our organization. Committed customers, not satisfied customers, are the stable base we need in order to grow. Committed customers, not merely satisfied customers, drive revenue, profit, productivity and every other concrete business measurement we use to gauge the health of our business. It's a blinding glimpse of the obvious — no customers, no results to measure.

"Because our plant has managed to integrate 'quality' into our everyday operations," Jim continued, "for us the answer is clearly 'yes,' but I would challenge each of you to answer for yourself the question, 'Is *my* quality initiative paying off?'"

How Will I Know if My Quality Initiative Is Paying Off?

Finding the answer to that question is the focus of the quest you began when you picked up *Quality on Trial*. Lack of an answer was what perplexed the CEO in Chapter 1 who could not link his organizations quality initiatives to bottom-line business results. He did not have an intuitive grasp of the concept of quality as a social movement. He did not understand that the ultimate goal of a social movement is to have its concepts and values integrated into the existing culture. Because the concept and value of customer commitment was not integrated into the ongoing business practices of his organization, for him, the jury remained out on the question, "Is my quality initiative paying off?"

Contrast the Chapter 1 CEO with the more visionary general manager, Jim. The latter realized that integration of quality occurs only when quality initiatives and business objectives are indivisibly linked. SPC, QFD, TQM and the like are not quality initiatives implemented for their own politically correct sake; they are means of achieving business objectives. And the *only* sure way of determining if they are, in fact, contributing to business objectives is by measuring their impact on your organization's relationships with its customers.

The Quality Review process establishes a clearly defined link between customer information and internal improvements; from information provided by customers, a company can determine whether SPC, QFD, specific attributes of TQM or any other "quality" initiative will help it strengthen its relationship with its customers. However, beyond front-end information collection that indicates "what should be done," the Quality Review process is a tracking mechanism for evaluating the impact of chosen actions and addressing "how well it is being done." There is only one way to evaluate the impact of a quality initiative in a business sense — talking to customers.

If a quality initiative is to pay off from a business perspective, there must be a clear link between internal improvements and the way in which the customer responds to those improvements. If the response is positive, the customer remains a customer, and there is a measurable impact on business results. The following examples from companies that have implemented the Quality Review process illustrate that series of linkages.

Business Result: Quantifying Employee Commitment

How important are employees to bottom-line business results? If employees are not committed to the vision of satisfying customers, then there is very little chance that customers will be satisfied or retained, or that the business will grow. But can those assumptions be quantified?

A large paper goods manufacturer would most assuredly answer "yes" based on its experience with TALK[2]. Responding to employee suggestions gathered through TALK[2] interviews, the managers, departments and plants took such actions as:

- Increased cross-training, which decreased overtime paid during vacations and absences and saved $3,000 per year.
- Developed a stringent project screening process to focus development efforts on projects with the highest probability of implementation, which reduced the number of projects processed by 18 percent and saved $12,000 per year.
- Revised the data entry program used to process contracts, which reduced contract processing time by 95 percent and saved more than 1,400 staff-hours and $28,000 in the first year.
- Eliminated duplicate reports and documentation through the use of an information transfer system, which saved 150 staff-hours and $12,000 the first year.

Remember — these improvements were just some of the changes made because managers took the time to meet individually with employees and apply the structured interview techniques of TALK[2]. Managers used their interviews to draw out improvement ideas and to work with each employee to refine the ideas into projects that could be implemented. This type of interaction does not occur in the standard, anonymous suggestion program. TALK[2] requires a little more work, but the evidence indicates that the bottom-line business results are worth it.

Business Result: Increased Market Share

A major oil company significantly increased its annual sales to domestic and international airline customers as a result of customer information gathered through the Quality Review process. The initial interviews of key contacts within the airline companies (who are the oil company's customers) identified requirements and expectations in a number of areas the customers considered critical to their businesses. Using the customer-defined requirements and expectations gained from the Quality Review process, the oil company management laid out a plan that addressed the newly defined requirements and expectations. *Taking advantage of the relationships developed during the initial Quality Review process interviews*, the oil company's representatives presented the plan to key individuals within the airline companies for their input prior to putting the plan in place.

Immediately following these meetings, the oil company received major fuel orders from two of their largest customers. In addition, every other customer participating in the Quality Review process indicated that they planned to do more business with the company in the future.

The Quality Review process is now integrated into the ongoing business practices of every functional unit of the oil company. Internal process improvements in areas like invoicing procedures, cash management policies and practices, and customer communication are driven by the information employees collect on a continuous basis — as part of ongoing business practices — from their customers.

Business Result: Gaining New Customers.

"We talk to customers all the time." This common refrain is heard often when the concept of the Quality Review process is introduced to organizations. And indeed it's true — companies talk to customers and potential customers all the time. The difference between traditional company-to-company communication and the Quality Review process is the kind and context of that communication. The purpose and the structured nature of the Quality Review process often lead to "discoveries" that are missed in traditional communication.

Take the example of a world leader in the processing of animal by-products. This organization implemented the Quality Review process in conjunction with pursuit of ISO 9000 certification. During a four-month period, company employees interviewed a large number of international companies — both current and potential users of their products. Employees talked to the usual contacts in purchasing as well as to people at the operations level. These operations contacts were outside both the normal sales process and the company's traditional range of contacts.

One operations-level contact was between the company president and the operations manager of a German plant. (In the past, the two companies had contact at the purchasing department level only.) This plant was not a customer; the plant believed the company's tallow did not meet the plant's specifications.

The new contact allowed for a new relationship, with new questions asked and new issues raised. The companies found that the tallow did meet requirements and was preferable to that of the German firm's current supplier. As a result, the company obtained a significant export order.

This example illustrates how changing or increasing the number of contacts within a customer (or potential customer) organization through the Quality Review process can lead to better understanding of needs, and result in significant revenue gain.

"World-class quality can only be defined by the customer — not chasing the competition, not focusing inside, not arrogantly assuming what the customer wants — but rather, getting the entire organization close to the customer, meeting the existing needs of the customer, and anticipating their future needs."

John J. McDevitt
Corporate Economist
3M Company

Business Result: New Product Development and Introduction

The results of the Quality Review process are not limited to sales and marketing organizations. By using the Quality Review process, the research and development (R&D) division of a large, diversified production and distribution company saved the corporation millions of dollars by directing their resources away from a developmental project they assumed would be of value to several of the company's large customers.

After several attempts to involve reluctant scientists and engineers in the Quality Review process, R&D management conducted several interviews with the prospective customers of the new product being designed. The information they received from these interviews indicated that the product development staff was not in tune with the needs of their customers. The information from the management interviews proved to be very convincing to the division's scientists and engineers. It caused them to re-evaluate their participation in the Quality Review process. Subsequently, many scientists and engineers became involved in interviewing their customers.

After completing the full Quality Review process cycle, the product development staff significantly altered their standard design, testing procedures, delivery schedule and documentation procedures. They established a regular customer contact schedule to ensure that their product was meeting customers' requirements. The Quality Review process is now fully integrated with the R&D division's new product development business practices.

Business Result: Customer Acquisition

A major communications company put the Quality Review process to the test. Company employees interviewed more than 2,200 customers. They also identified a comparable group of customers who were not interviewed. At the end of the test period, the cancellation rate for the interviewed customers was 49 percent lower than the rate for the control group.

In pure business terms, the lower cancellation rate resulted in greater revenues for the company. In addition, the company saved sales expense dollars because the revenue increase came from existing, rather than new, customers. Company surveys showed that it is six times more expensive to acquire a new customer than to retain and expand the business of an existing customer.

As a result of the success of the Quality Review process, the communications company has put in place ambitious plans for using it as part of a customer *acquisition* initiative. As part of their standard presentations to potential customers, sales reps share Quality Review information that shows how customers have identified "quality of service" as clearly separating the company from its competition. In addition, sales reps explain the Quality Review process to prospective clients as the *process* used to deliver that superior level of service — a tactic that distinguishes the company from others that merely *claim* superior service. In other words, the Quality Review process proves that the company can deliver on its service promises.

The demonstrated business value of the Quality Review process to this communications company is such that it is fully integrated into the company's sales organization. All new sales reps learn how to conduct Quality Reviews as part of basic sales training. Each sales rep conducts between one and four Quality Reviews per month. Plus, conducting Quality Review interviews is part of the performance criteria for induction into the company's sales recognition programs.

Business Result: Customer Retention

The Quality Review process has proven to be a dynamic process. It is not necessary to complete the entire Quality Review cycle before taking actions that have significant impact on business results. A large manufacturer of semiconductors teamed members of their R&D, engineering design, and sales and marketing staffs and chartered them to interview multiple contacts within 50 of their largest customer organizations. Within a matter of a few days, the review teams discovered that three of their major customers were close to defecting to the competition. To address this problem, the president formed two task teams, which developed specific action plans for each customer. These action plans were immediately implemented. Each of the potential defectors subsequently expressed a

Business Result: Employee Morale and Retention

The value of the Quality Review process lies in its ability to direct the activities of a company's quality initiative toward those processes that will have a high impact on customers and consequently a high impact on business results. A pleasant side benefit, however, is the impact of customer contact on a company's employees.

One of the most outstanding results reported by companies using the Quality Review process is that employees are learning the most important ingredients in any quality initiative — who their customers are and how to systematically determine their requirements and expectations and anticipate their needs. This external perspective, the notion of serving customers backed by a concrete methodology for putting that philosophy into action, gives a new meaning and purpose to people's daily jobs. A sales coordinator of a health care division of a major corporation provides evidence to this point:

The Quality Review interview process was a REAL growth experience for me, and helped to broaden my perspective in several ways.

It increased my awareness of the important role our division sales management team plays in building and maintaining good distributor relationships. Just one example of this is the work management does with distributor management to develop meaningful incentive programs. There were a number of other examples.

Secondly, it was an "upper" for me to hear first-hand from our customers of the high regard they have for our company, both because of our outstanding product quality, their perceived image of our integrity as a corporation, and because of the high caliber of people they deal with in our company.

*Third, having the opportunity to hear **first-hand** from our customers what they would like us to do better or differently will make me a committed advocate of change in these areas. On the other hand, I will be a staunch supporter of status quo in the areas our customers feel we do well.*

*Being a part of the Quality Review interview process is the **most exciting** thing that has happened to me in my 10 years with my company!! (And, by the way, I have enjoyed every position I have held here.) I hope that I have the opportunity to be a part of the ongoing Quality Review process, both in the training process for additional people brought in and in the ongoing interview process with distributors and with hospital and end-user customers.*

In addition to generating business results through applications with external customers, the Quality Review process is establishing an impressive track strengthening the relationships among departments and work groups within an organization.

renewed commitment to the company. And, all three customers expressed their appreciation that the company had involved its employees (rather than a third party) in the process of gathering and acting on information critical to their [the customers'] businesses.

Business Result: Gaining Additional Business from Existing Customers

In their book *The One-to-One Future: Building Relationships One Customer at a Time,* authors Peppers and Rogers identify a trend they call the "miniaturization of marketing." They note that successful companies in the '90s will build market share not by increasing their number of customers, but rather by selling more of their products to existing customers.[2] This kind of focused selling requires the in-depth customer knowledge that the Quality Review process is designed to deliver.

Take the case of a large, international engineering consulting firm. Using the Quality Review process, the engineering firm conducted more than 20 interviews with a single customer organization. The interviews focused on a specific project. Armed with information gathered through the process, the engineering firm addressed customer issues that were affecting both the implementation of the project and the customer's commitment to the firm.

Around the time the interviews were taking place, the customer organization was planning two additional projects with combined consulting fees of about $20 million. The engineering firm knew of these projects, but felt it unlikely that it would be awarded either project. However, the customer said the engineering firm had demonstrated its understanding of the company's project needs and requirements — and then awarded the firm one project outright. The second project was won through a bidding process. The firm used the Quality Review process to develop the kind of relationship that allowed it to capture a larger share of this customer's business.

Business Result: Cost-Effective Use of Resources

As part of its strategic planning process, the Management Information Systems (MIS) department of a large, diversified computer manufacturing company implemented the Quality Review process as a means of identifying internal customer requirements. This step represented a significant change in both operating procedure for the organization and in its orientation to the rest of the company. The MIS group had always played the role of the "expert" who knew what systems were best for end users. Implementation of the Quality Review process sent a message to the organization that the MIS department was adopting a more partnership-oriented attitude.

Following the steps of the Quality Review process, the organization formed a team to identify key quality indicators. Employees from all the functional groups within MIS were trained in interviewing customers. Customers were identified. A communication effort was mounted to inform customers of the MIS department's new approach to strategic planning. The interviews were conducted.

As a result of implementing the Quality Review process, the MIS department got a very real-time, realistic view of its service to the organization. Customers — users of MIS services — told the department that it did not rate well when it came to accessibility. Phone calls were not returned promptly, voice mail was not used effectively, and the hours of operation (8 a.m. to 5 p.m. Central Time) were not convenient to the company's East and West Coast facilities. When asked to evaluate MIS projects, customers said the department programmers and systems people were working on projects that had low priority for them. MIS users wanted more PC and network support; at the same time, MIS management was considering the purchase of a larger mainframe to handle more application work.

As a result of this data, the MIS department took several actions. Some changes were major, such as scrapping plans for the larger mainframe and redirecting efforts to PC and network support, and setting new priorities for programmers that were more in line with the requirements of the MIS users. Other changes were very simple, such as providing beepers to key people so they could return phone calls in a timely fashion, and setting up a split shift so that the MIS center was staffed to provide a full day's coverage for East and West Coast facilities.

Business Result: Cost Reductions

Organizations who have carefully managed the actions they have taken in response to their internal customer requirements and expectations have realized significant business results. One manufacturing company, for example, after completing their first full cycle of the Quality Review process, documented cost reductions in excess of $10 million. These reductions in cost were realized as a result of a joint effort on the part of managers and employees (representing different departments) planning and implementing actions in response to the direct feedback they had obtained from their customer-supplier interviews.

"Is your quality initiative paying off?" Having passed through the Evaluative Component of a quality movement, one would imagine he or she is now in a position to answer that question. But, and this is indeed an irony of the quality journey, after integrating the five components of a quality movement — Visionary, Formulative, Rhetorical, Activist and Evaluative — the question that started the quest is no longer the right question.

Recall the comment of the visionary general manager in the example that opened this chapter: "We don't talk that much about *quality* processes. We're usually talking about our *business* objectives. For us, quality and business objectives are indivisibly linked." Thus, when quality is fully integrated into an organization, the only valid question is, "Are we achieving our business objectives?" Quality initiatives, like R&D projects, like marketing strategies, like sales motions, like operations productivity improvements, are elements that contribute to business objectives. The integrating factor is customer information — information about customer requirements, expectations and needs — gathered by an organization's employees through direct contact with customers.

However, in today's rapidly changing business environment, consecutive short-term successes alone are not the way to "run the railroad" for long-term success. Despite the impressive immediate business results achieved through use of the Quality Review process, we contend that the real payoff occurs when the process is totally integrated into the standard operating procedures of the organization. How is that accomplished? We examine that question in Chapter 10.

---

[1]   Management by Objective

[2]   Don Peppers and Martha Rogers. *The One-to-One Future: Building Relationships One Customer at a Time.* Copyright 1993 Doubleday, New York, NY.

"As the leader of your organization's quality initiative, your job is to get business and quality dancing to the beat of customer information."

10
Achieving Integration

"WE HAD OUR BASIC MANAGEMENT SYSTEMS IN PLACE," SAID GENERAL manager Jim in the previous chapter. "If we were going to *achieve* quality," he added, "we had to *integrate* quality processes and customer information into those existing systems."

Recall from the discussion in Chapter 4 that a vision that is meaningful to an organization is both a departure from and *an extension of* the status quo. When Jim's organization embarked on its quality initiative, it did not do so by totally disrupting its organization. Rather, it integrated new ideas and new ways of working (including a customer management system) into existing business processes. The result was a quality initiative that was "just the way we work" — not a parallel activity to the "real work" of the organization.

As the customer information management system underlying a quality initiative, the Quality Review process should also be integrated into ongoing business practices. Full integration of the Quality Review process is achieved only when the collection and use of customer information becomes "real work" — part and parcel of the way an organization does business. Customer information must be

Sometimes You Gotta Raise Some Blisters

FORT MEYERS, FLA. — Tuesday morning the [Minnesota] Twins showed why they won the World Series last October.

[Manager] Tom Kelly took his catchers and infielders and hit them ground balls for 75 minutes in an annual drill he calls "rocket fire." During the drill, Kelly hits grounder after grounder — every six seconds in some stretches — while fielders execute various game scenarios.

Kelly wore out one fungo bat and raised blisters on his hands hitting at least 600 grounders during the drill.

"You hope it makes an impression," he said. "That's why I do it myself. When the *manager* does it, it tells people this must be important if the manager does it. I could sit on the sidelines and have a coach do it, but I want to make an impression."

St. Paul Pioneer Press
March 4, 1992

integrated into both internal and external business practices, and ultimately into the fiber of strategic planning. Similarly, if employees are to be integral to the process of building customer commitment, their views must also be integrated into an organization's business practices through a process like TALK^2.

Integration is the leader's responsibility. Without being prescriptive and dictating the tactical methods of integration to the various departments within an organization, a business executive must set the strategy for company-wide integration of customer and employee information. He or she must be an active force in that integration. Based on our experience with companies implementing the Quality Review and TALK^2 processes, we offer the following suggestions.

Making Customer Information Available to Everyone

To effectively integrate customer information into the strategic-planning process, managers and other selected employee groups at multiple levels in the organization must have ready access to both the information that has been collected and analyzed from customers and the actions that have been initiated to respond to customer needs. An integral part of the Quality Review process is the customer management information system, developed by Quality Institute. This system makes the information collected from customers available to managers, professionals and technical personnel via computer terminals or personal computers at their desks or workstations.

The system enables users to examine information such as:

- Results of individual customer reviews, including both qualitative and quantitative data.

- Customer Confidence Index scores by customer or customer group.

- Aggregated customer data for a specific region or product line, by volume of purchase, or by any other selected factor.

- The number of customers reviewed, when they were reviewed, and by whom.

- Comparison information against competitors or industry trends.

- Improvement trends for individual customers or customer groups.

- Actions taken to respond to individual or collective customer needs.

- Business results achieved from actions taken.

Customer information, continuously gathered and acted upon during the Quality Review process, should be included along with financial, marketing and manufacturing data for both strategic and operational decision making.

Incorporating customer information into the management reporting structure sends a clear signal to the organization that such information is integral to the way the company does business. It provides continuous exposure to customer information and stimulates executive management to ask questions they never addressed before.

Integrating Customer and Employee Information into Strategic Planning

Ultimately, integration of the Quality Review process and TALK[2] into internal and external business practices should come together in the strategic-planning activities of the organization. Relationship management based on information gathered from customers and employees becomes the driving force behind the company's strategic plan. Successful integration of relationship management into strategic-planning activities is a sure sign that your organization has come to realize the value of customer relationships and is positioned to exploit that knowledge for competitive advantage.

Top-Line Findings

In many businesses, senior management does not review quality performance even monthly. In contrast, most financial reviews occur more frequently, and 10-day sales reports are not uncommon.

Until the reporting of quality results achieves parity with traditional financial and operating reporting, the link between quality, performance and profit cannot be fully realized.

Too many North American and German businesses do not yet have a process designated to measure explicitly the business consequences of quality performance.

If businesses are to drive change in their quality-related practices and strategic direction, they must develop measurement systems that link quality to performance and then make such systems an integral part of the management process.

American Quality Foundation and Ernst & Young International Quality Study, 1991

"The need to computerize quality exists across the nation: from billion dollar companies to small suppliers; from aerospace manufacturers to banks.... Overwhelmingly, the Customer Feedback System is the highest priority for computer assistance."

Quality Management
Forum Survey, 1991

Traditionally, strategic planning has integrated the organization's business goals with the activities of the marketing, sales, R&D, operations, and support and administrative functions. In a top-down process, goals are set and then activities defined that will achieve those goals. All activities of the organization are integrated to achieve the goals of the business.

The Quality Review process and TALK[2] facilitate the integration of customer and employee information into that equation. However, the introduction takes place *within* the organization, not at the top. The requirements, expectations and needs of customers are driven up the organization. Action plans that address those requirements, expectations and needs are also fed upward. When appropriate, the top of the organization then provides the resources necessary to achieve the action plans. Inherent in this methodology is recognition that unless the requirements, expectations and needs of the customer are met, business goals cannot be achieved.

Integrating Customer and Employee Information into Operations

Operations functions, especially manufacturing, have been among the prime targets for quality initiatives. Generally these initiatives have taken the form of process improvement, focusing on preventing defects as a means of increasing productivity, reducing latent product defects and infant mortality in the field, and reducing cycle times. Reduced defects and cycle times (shorter lead times, from the customers' perspective) are benefits that may contribute to customer satisfaction and reduced internal costs. However, as indicated in previous chapters, internal improvements in and of themselves do not necessarily contribute to customer commitment.

Integrating customer and employee information into operations allows the organization to focus improvement of internal processes on those areas with the greatest impact on the customer (which may or may not be those with the most internal impact). When this integration is accomplished, all quality initiatives will be evaluated on their impact on the processes critical to meeting customers' requirements, expectations and needs.

*Integrating Customer Information
into Marketing and Sales*

Typically, the marketing and sales group is the primary source of customer information in an organization. Customer information flows in one direction — from marketing and sales to the other departments in the company. The by-product of this flow is the filtering of customer information by marketing and sales. Whether this filtering is intentional is irrelevant. The point is, customer information is interpreted for the organization from a single limited perspective.

The Quality Review process differs from the traditional approach in that customer information flows into the organization through all departments. When development and operations personnel have access to the unfiltered voice of the customer, they can better improve product and service quality *from the true perspective of the customer.* On the other hand, if the people in marketing and sales predigest the information, opportunities may be missed because marketing and sales people often do not have the technical knowledge to connect user needs with innovative solutions.[1]

The Quality Review process allows for passage of customer information not just from marketing and sales into the organization, but from inside the organization to marketing and sales. The value of this point should not be overlooked. Internal departments, armed with direct information from customers, can and do provide unique insight into customer "hot buttons" that aid in the sales motion.

When the Quality Review process is fully integrated into the way a company does business, information from many customer sources — both purchasers and purchase influencers — is available to the company. Integration of customer information means that all organizations can and should participate in strategic and even tactical account planning.

> ### Customer-Focused Operations
>
> The rules have changed. People skills, marketing skills and customer-relations skills are now critical for *manufacturing* executives, because they spend a great deal of time in the external world outside manufacturing — particularly with customers.
>
> *Operations* VPs now help start or close a deal, assure an edgy customer that the product will be delivered on time and often handle product complaints.
>
> J. Mirtz
> *Industry Week*, October, 1989

"To be of greatest use, customer information must move beyond the market research, sales, and marketing functions and 'permeate every corporate function' — the R&D scientists and engineers, the manufacturing people, and the field-service specialists."

B. Shapiro
Harvard Business Review,
November/December, 1988

Other ways in which integration of the Quality Review process into external business practices is evident include:

- Internal organizations are indirect participants in the preparation of promotional and advertising materials. Departments that actually design and build products (or deliver services), and that have interacted with users of the products or services through the Quality Review process, bring a perspective that traditional customer contact organizations might miss.

- Internal organizations are actively involved with marketing and sales in ongoing external communications with customers. This activity capitalizes on relationships built through Quality Review interviews and demonstrates the desire of the organization to continue those relationships. With strong customer relationships come opportunities to "up-sell" existing accounts and for customer referrals for new account development.

- The idea of customer relationships as corporate assets is communicated to shareholders and the financial community. Information gathered from customer interviews and the analysis achieved through utilization of the Customer Confidence Index can complement financial measures as a way to assess the overall health of the company.

Integrating Customer Information into Research and Development

The ability to anticipate customer needs is the cornerstone of creating committed customers. Most companies travel a circuitous path to develop new products — the marketing department makes a product requirement suggestion and the R&D department responds, or R&D initiates product development based on some nifty new technology that it is sure customers will want. Either case reflects the old "throw it over the wall to the next department" syndrome that flourishes in many companies. In many organizations, the R&D department is not exposed directly to the customer.

This lack of exposure blocks the organization from its primary source of innovative ideas. Studies confirm that the lion's share of new product ideas come from users,[2] yet according to the American Quality Foundation, fewer than 25 percent of U.S. companies translate customer expectations into the design of new products or services.[3] It is the interaction with these users (customers) that will ultimately make possible the anticipation (not just meeting) of their needs.

Effectively integrating customer information into the R&D function begins with leadership clearly establishing that the role of R&D is to create customer commitment. This commitment is based on the ability of its employees to create products that respond to customer requirements, expectations and future needs. To further achieve integration, the leader can:

- Position the integration of the Quality Review process as a way to increase the involvement of R&D employees with customers to enable a better understanding of customers' requirements, expectations and future needs.
- Stress the need to make building customer relationships a part of the R&D reward and recognition system, equal in status to technical achievement.
- Explain the need to use the Quality Review process with the internal customer groups that are impacted by R&D activities (such as sales training groups, support services, and the like).

Integration of the concept and value of customer commitment into an R&D department means that value to the customer will always be the guiding principle of the group (Chapter 3).

Integrating Customer Information into Human Resource Management

The human resource function can play a critical role in helping all employees understand that the mission of the organization is meeting customer requirements and exceeding customer expectations, and that the vision of the organization is creating committed customers. Human resource leadership can ensure that every human resource policy is matched to this standard. Accordingly, the function can pursue the following actions as appropriate:

- Incorporate into reward and recognition programs criteria that clearly reinforce the external focus of the organization.
- Incorporate in the job descriptions for both work groups and specific individuals the employee's responsibility for effective relationship building with customers.

"The only good news ... from this global study is that 51 percent of U.S. companies plan to use their company's quality performance as a primary criterion for compensating senior management, up from 19 percent that currently do so."

Joshua Hammond, President
American Quality Foundation

Integrating the Customer Confidence Index™

Quality Institute's Customer Confidence Index provides information to work groups and action-planning teams, helping them direct their focus and identify priority targets for improvement (Chapter 6). It ranks those quality indicators in which customers have the least and the most confidence. The index also identifies which customers have low, moderate or high levels of confidence in the organization's performance. It identifies customer relationships that may be at risk, allowing the organization to take corrective action with the individual customer before the customer relationship is lost.

Those are the primary functions of the Customer Confidence Index. However, like other aspects of the Quality Review process, the greater the integration of the Customer Confidence Index, the greater the impact on the organization. Other ways to integrate the Customer Confidence Index into ongoing business practices include:

• Use the Customer Confidence Index to identify the outstanding performance of work groups or individuals who develop and maintain excellent customer relationships. Recog-

nition is based on index scores reflecting direct customer feedback.

• Add the Customer Confidence Index to the list of factors currently used in determining management performance. The index provides direct customer feedback describing the effectiveness of the work group or organization under the responsibility of the manager. It is a direct measure of an organization's (or its work groups') commitment to customers.

• Make trending of the organization's performance and commitment to customers as tracked by the Customer Confidence Index part of the management reporting system. Comparisons can be made between groups, industries and national norms as appropriate.

• Incorporate information provided by the Customer Confidence Index into the organization's strategic planning process. The index provides customer feedback about the organization's strengths and opportunities for improvement. The index scores can help management determine where an organization needs to focus its attention.

• Establish clear customer-related criteria as part of every employee's performance evaluation.

• Establish customer commitment as a consideration for merit pay increases, incentive pay and promotion considerations.

• Communicate in the employee handbook and/or other employee publications the expectation that employees participate in the collection of customer information.

• Use the Quality Review process to evaluate human resource initiatives with the function's internal customers.

The principles of the company's mission and vision must be integrated into the basic philosophy, policies and systems of the human resource function.

Integrating Customer Information into Employee Training

The principles of the Quality Review process can be integrated into an organization's training initiative in four ways:

- Employees are trained to conduct customer interviews and to bring the knowledge they gained back into the organization.
- Sales personnel are trained to use the Quality Review process as a relationship management strategy.
- Information gathered from customers is used to determine what training needs exist, such as who needs to be trained and in what areas.
- Current training initiatives are evaluated using the Quality Review process with internal customers, e.g., employees who receive training.

Customer information gathered through the Quality Review process is the basis for decisions about what and who should be taught. Training must be related specifically to customers' needs. Training facilitators need to ask themselves, "What is the link between the training we are providing and building customer commitment? How will implementation of what is taught contribute to building stronger relationships with customers?"

Integrating Customer Information into Supplier Relations

It is a rare organization today that does not have a formal supply-line management program in place. As discussed in Chapter 8, these programs are typically focused on managing the flow of material into the company, not on managing the relationship with the supplier. To achieve integration, leaders need to emphasize to employees that:

- Supplier conformance to requirements is only one sign of the health of the relationship between the supplier and the organization; the purpose of the supplier Quality Review process is to expand that relationship by increasing the number of employees involved with suppliers.
- The ultimate objective of the supplier review process is motivating the supplier to conduct its own Quality Review interviews with the organization.

- The procurement organization can use the Quality Review process with its internal customers to monitor its own performance and identify how the purchasing function can best serve the organization.

When your organization conducts supplier reviews, it motivates your suppliers to conform to company requirements so your company can conform to the requirements of its customers. When fully integrated into ongoing business practices, the supplier Quality Review process enables suppliers to treat your customers as their own (Chapter 8).

As our research reveals, one of the major failures of most quality initiatives is the perception of employees that quality is parallel to the real business of the business. Quality never becomes integrated into the daily activities of the organization — business advocates remain on one side of the ledger, quality gurus on the other, glancing nervously at each other like boys and girls at their first junior high school dance.

Quite simply, as the leader of your organization's quality initiative, your job is to get business and quality dancing to the beat of customer information. Your instrument is the Quality Review process. The customer information provided by the Quality Review process, when acted upon at all levels of the organization, will produce business results for your company like those discussed in the examples in Chapter 9.

So then, get the organization acting on customer information, and your quality initiative is a success, right? Recall from Chapter 2 that the quality movement is a social movement and that the ultimate goal of a social movement is to have its concepts and values integrated into the culture of the organization. Short-term results should not be the end-state goal of your quality initiative. Gathering and acting on customer information is only a partial success. Ultimately, it is integration of *the system* of gathering and acting on customer information that is the litmus test of success. This is not mere semantics.

Nobody bats a thousand. No organization will always correctly interpret customer feedback, nor will it always take appropriate action on correctly understood customer desires. But, an organization that is focused on the process of gathering customer information will take advantage of its relationships with its customers to understand its failures. An organization with its focus *only* on immediate results will throw out the guru with the bath water and go looking for another panacea. We have seen that scenario played out over and over again — companies bouncing from one quality initiative to another without ever asking their customers why the organization is missing the mark.

In this chapter we've introduced some management behaviors that will help an organization integrate the system of gathering customer information into the company through the Quality Review process. Do not underestimate the importance of that task. The case for importance has been made and the verdict is in: Integrate customer information into ongoing business activities, and quality pays. But just how difficult is that task?

As is the case with most undertakings, the more you understand about changing people's behavior, the easier it becomes. Chapter 11 looks at the Quality Review process as a new paradigm of customer management. The concepts presented in Chapter 11 will help you answer objections to the Quality Review process, focus your efforts on behavior change, and perhaps even address some of those nagging objections to the Quality Review process as it challenges some of your own cherished beliefs.

We've been there. We've raised the objections. We've wrestled with the anxiety and stress of changing the way we understood customer measurement and how we acted on it. We changed our behavior toward our clients and toward their customers. Chapter 11 is our map of the territory. It's a land worth exploring.

---

[1]   Benson P. Shapiro. "What the Hell Is 'Market Oriented'?" *Harvard Business Review,* November/December, 1988.

[2]   Tom Peters. "The Mythology of Innovation," in J. William Pfeiffer's *Strategic Planning.* Copyright 1986 Pfeiffer & Co., San Diego, CA.

[3]   American Quality Foundation/Ernst & Young International Quality Study. Copyright 1991 American Quality Foundation and Ernst & Young.

"The new paradigm allows an organization to move up the Mission/Vision Hierarchy from merely satisfying customers to developing committed customers by meeting, exceeding, anticipating and addressing their previously unrecognized needs."

11
Throwing Down the Gauntlet of Change

PLAYING CARDS, ONE AT A TIME, ARE RAPIDLY FLASHED ON A SCREEN before you. Five of hearts. Queen of clubs. Ten of diamonds. Three of spades. Ace of clubs. You have no trouble identifying the cards as rapidly as they appear. But as the speed at which the cards are shown slows, you get an uneasy feeling. That last card — it was the three of spades, wasn't it? It looked like the three of spades. It must be the three of spades.

The rate of exposure of the cards slows a little more. Your uneasiness increases. You can still identify most of the cards at a glance, but some, for no explicable reason, give you trouble. Three of hearts ... no, spades. Three of spades. Yes, definitely the three of spades.

Slower yet. Five of hearts. (No problem.) Queen of clubs. (No problem.) Ten of diamonds. (Ditto.) Three of ... suddenly you realize the cause of your consternation. The card is indeed the three of spades, *but the spades are printed, not in the traditional black, but in red.* As the cards continue to flash you notice other anomalies — a black nine of hearts, a red Jack of clubs, a black seven of diamonds. Even after understanding that the suits on some cards have been altered, you still can't shake a somewhat uncomfortable feeling when looking at the mis-colored symbols.

The above scenario is, in fact, an experiment described in Thomas Kuhn's 1962 work, *The Structure of Scientific Revolutions.*[1] The significance of the experiment to our discussion of customer information collection, measurement of customer commitment and the Quality Review process has to do with the concept of "paradigms."

In simple terms, the paradigm concept presumes the notion that people tend to organize their thinking in familiar patterns. Those patterns, or paradigms, govern the way people interpret data. In the playing card experiment, when people were briefly exposed to a red three of spades, a black nine of hearts, and so on, they invariably identified the cards by the pattern with which they were most familiar. The red three of spades was seen as a black three of spades or a red three of hearts, the black nine of hearts as a nine of hearts or a nine of spades. Significantly, even after exposure time was slowed to a point where people noted something was amiss, *they continued to identify the anomalous cards as if they were part of the standard deck.* In other words, the people in the experiment continued to interpret the data they were seeing (anomalous cards) within a familiar pattern or paradigm (the standard 52-card deck).

Treated with intellectual rigor in Kuhn's work, the "paradigm" and corollary "paradigm shift" concepts have been rediscovered by the business world and made the unfortunate transition from cutting-edge concepts to business buzzwords. If one accepts at face value the rhetoric whirling around Qualityland these days, one might (erroneously) assume that shifting a paradigm is as simple as changing the corporate advertising slogan. In today's business lexicon, "paradigm" and "paradigm shift" are two of the more broadly misused terms. In the minds of many business leaders they are associated with the *style* of the quality movement rather than the *substance* of the movement. That's too bad, because once one strips away the hyperbole and hype, "paradigm" and "paradigm shift" are useful concepts for understanding and managing change.

"Incrementally fixing the old, broken bureaucracy just isn't doing the job these days. What's required are quantum ideas for products and services, as well as revolutionary changes in the organization to produce them."

Noel Tichy
Professor of Business
University of Michigan

The Quality Review Process as Paradigm Shift

Although not a paradigm shift of Darwinian or Copernican impact, the paradigm of customer measurement, we submit, has shifted from a macro-level, Ferris-wheel view of customers to a micro-level, relationship management approach. The Quality Review process model of customer measurement may be but page-two news in the cosmic order of things, but it nonetheless represents a substantial pragmatic change that will significantly alter the thinking and behavior of the business community. And by Kuhn's criteria, the Quality Review process, within the scale of the business community, is indeed a true paradigm shift.

This book throws down the gauntlet of change. It does so because change is no longer up for election. Change is *the* business survival issue of the '90s. What should be changed? *Quality on Trial* has challenged thinking on four key points:

- Quality as a free-standing, isolated set of activities is a luxury that cost-conscious companies can no longer afford. Quality must carry its own financial weight as an integrated part of the way a company does business. Quality for quality's sake is not economically justifiable.

- An organizational mission of creating satisfied customers is no longer good enough. In order to grow (and survive) a company must move to the more visionary approach of creating customer commitment.

- The driving force of the quality movement must shift from an internally driven, inside-out focus on process improvement to an externally driven, outside-in focus on the requirements, expectations and future needs of the customer.

- The view of the customer must move from the detached Ferris wheel perspective that interprets customer information on a macro level to a relationship management perspective that interprets customer information on a customer-by-customer basis.

Traditional market research and third-party customer measurement techniques are proving inadequate in the new business environment. They may provide *necessary* information for managing a business, but they do not provide *sufficient* information for the

What's Wrong with Traditional Research?

Many American companies have introduced customer satisfaction and service systems during the past decade.... Some of the work on measurement is good. Much, however, is problematic. Among the more common problems are:

- *Using mailed questionnaires.* These yield a selective sample that is not representative of a firm's customer base. [People] like to cheer and complain. Mailed satisfaction studies produce disproportionate numbers of both cheerleaders and gripers. The middle range of reasonably contented or mildly displeased customers fail to cooperate and are therefore not included in the resultant database.

- *Ignoring customers' needs.* A surprising number of studies are done in which customers are asked to rate markets and competitors on a variety of dimensions without recognizing (and measuring) that these dimensions may motivate individual customers differently. Some people may be price-conscious, others quality-conscious, still others service-oriented. All dimensions of evaluation, in other words, are not created equal, and it does not take very sophisticated technology to figure out which ones are more important than others.

- *Conducting death-wish research.* Death-wish research includes all of the flaky research tools discussed earlier when we talked about positioning and advertising. You don't measure customer satisfaction and service quality through the use of focus groups, shopping-mall samples, user conferences, importance ratings, gap analyses, and the like, unless *you want to reach the wrong conclusions and make the wrong decisions.*

K. Clancy and R. Shulman
The Marketing Revolution, 1991

demands of today's business environment. In Kuhn's terminology, the old paradigm leaves business leaders facing "persistent anomalies" — persistent contradictions between collected data and actual business results; for example, data showing happy customers in the face of declining sales.

Our conversations with hundreds of CEOs, quality professionals, and line managers at many corporate levels indicate that executives are intuitively recognizing that the paradigm is shifting from customer satisfaction to customer commitment. They are recognizing the need to look outside, not inside, their organizations for the *raison d' être* of their quality initiatives. Executives are putting quality on notice that it must contribute to bottom-line business results. They are realizing that the question "Is my quality initiative paying off?" is no longer the relevant question. The question that must be addressed is, "Is my quality initiative contributing to the achievement of our business objectives?"

Because the Quality Review process is based on the concept of an organization's employees — people with intimate knowledge of the organization's business interviewing customers — a greater level of detail (than third-party collected data) about customer requirements and expectations is a natural outgrowth of the effort. Information gathered on a customer-by-customer basis through the Quality Review process can identify which of a company's quality initiatives are impacting the customer, hence which ones are having the greatest impact on business results, hence which ones are paying off. Further, the Quality Review process provides indicators as to what additional business initiatives (based on customers' stated requirements and expectations) ought to be undertaken, i.e., which new quality initiatives will impact customers, hence business results.

The ability of the Quality Review process to answer "crisis" questions such as these is important, but as Kuhn notes, it is but one criteria by which to evaluate a new paradigm. The second of Kuhn's criteria is the ability of the new model to predict phenomena that weren't even recognized under the old paradigm. In a business context, the new paradigm allows an organization to find solutions for problems that under the old paradigm *weren't even recognized as problems*. In our terms, the new paradigm allows an organization to move up the Mission/Vision Hierarchy from merely satisfying customers to developing committed customers by anticipating and addressing their previously unrecognized needs.

This ability results from the relationship management aspects of the Quality Review process. As the relationship between a supplier organization and its customers diversifies and becomes more intimate (Chapter 5), more information exchange occurs. The information flow from an organization's customers to its employees becomes seed for business growth as the organization's employees learn to anticipate and fulfill customers' needs.

The New Customer Measurement Paradigm

The currently practiced paradigm of customer satisfaction measurement holds that the primary purpose of customer satisfaction research is to provide management with data reflecting how customers generally feel about supplier organizations. However, as we have personally experienced, while traditional survey methods provide an organization with valuable data, the third-party methodology has the inherent disadvantage of not facilitating what marketing consultant Regis McKenna refers to as the "qualitative approach to marketing." [2]

A statistical study might show that 10 percent of a supplier's customers are dissatisfied, McKenna notes, but that piece of data alone is insufficient. It does not reveal qualitative information like the *intensity* of the dissatisfaction or if customers have reached a point where they are expressing their dissatisfaction to others,[3] nor does it define the relative *importance* of areas of dissatisfaction to the customer or whether the supplier is *improving* or regressing — the key elements of the Customer Confidence Index (Chapter 6).

Despite the inability to provide qualitative data on a customer- by-customer basis, but consistent with a paradigm mindset, the business community is only just beginning to question the tradi- tional model of customer satisfaction measurement. Still very much standard operating procedure is the refining of customer sampling techniques, the modifying of survey questionnaires, and tinkering with the scope of a customer satisfaction survey — but throughout this face-lifting procedure, the underlying assumption that the purpose of high-level customer measurement is to provide quanti- tative assessment of a relationship still is rarely challenged.

The Quality Review process, on the other hand, is based on the premise that the primary objective of customer measurement is not just the gathering of quantitative customer information, but also analyzing qualitative aspects of the customer/supplier relationship and enhancing that relationship. In the language of Harvard mar- keting professor Theodore Levitt, "the era of the one-night stand [in supplier/customer relationships] is over." Today, says Levitt, the relationship is more like a marriage, where each partner is commit- ted to the other.[4]

Make no mistake about it — the Quality Review process is a radical departure from traditional customer measurement and customer information collection methods. And change, especially radical change, means resistance. Resistance to change takes two forms. The first is holding on to the old allegiance. The second is actively raising objections to the new. Executives who have implemented the Quality Review process are not strangers to either.

Old allegiances die hard. Recall the playing card experiment: even after sensing something was happening, people participating in the experiment continued to identify the mis-colored cards as if they were part of a standard deck. In essence, they were denying the sensory data they were experiencing and identifying the new cards within the context of the old paradigm. If merely changing the color of the shapes on a deck of playing cards elicits that strong a reaction, what are the expectations when an organization tries to change practices in which people have invested their egos and that are foundational to their view of the world?

Change of paradigm-shifting magnitude is not easy, as we have learned firsthand. Developing the Quality Review process challenged us to question some basic assumptions of "good research" that we had spent our careers studying, refining and implementing:

- **"Good Research" Assumption #1:** Customers have to be assured that their responses are anonymous and the information confidential to ensure that their responses will be candid and the data valid.
- **"Good Research" Assumption #2:** In order to collect unbiased data, the person collecting the data must be an objective third party.
- **"Good Research" Assumption #3:** The data collected from customers must be collected under controlled circumstances to ensure its reliability.

Why did we (and why does a business leader need to) challenge these assumptions? Each in its own way hinders organizations from using customer data and customer data collection to their maximum potential — the managing of customer relationships to build customer commitment.

Assumption #1: Ensuring confidentiality.

If customer information is not identified with its source, an organization loses the opportunity to respond to an individual customer's concern. This is one of the most powerful uses of customer data — to develop individual customer strategies.

In addition, our experience and research indicates that the vast majority of customers are more likely to provide thoughtful, candid responses to employees of the organization who is sponsoring the research than they are to an anonymous survey form or telephone

Can Sales Personnel Collect Good Data?

In the Quality Review process, employees who manage relationships with their customers (e.g., sales representatives) are asked to gather information from them regarding the organization's performance. The conventional wisdom says sending sales people in to gather customer data is like putting the fox in charge of the chicken coop, or that it sets up a fishing expedition for compliments. Are customers candid in their responses when talking directly with their key contacts? Are they likely to respond differently than they would if their key contact was not present?

To provide answers to these questions, Quality Institute conducted an empirical test comparing the results of customers' quantitative responses to questions when their sales representative was present to when their sales representative was not present. The analysis revealed no statistically significant differences between customer responses on any factors, including attitude of personnel, competence of personnel, sales support, product quality, accessibility, problem solving, communication, responsiveness, delivery, order processing, partnership approach and profitability. In other words, customers were equally candid in both situations.

Further, our experience with training and debriefing employees who conduct customer interviews continues to validate the ability of a wide variety of employees to effectively gather actionable customer data and, most importantly, to benefit significantly from hearing the voice of the customer firsthand.

interview. The last thing a customer wants to be is anonymous. A customer with an opinion wants to be heard. A customer with a problem wants that problem solved, not aggregated, analyzed, prioritized and sanitized into a generic solution. A customer wants evidence that his or her concern will be addressed, and when customer contact is made by a line employee of the supplier, customers believe the information they provide will actually be used to their benefit. They have far more confidence in action, far less fear that their comments will disappear into a bureaucratic black hole.

Assumption #2: Third-party data collection.

Gathering unbiased information remains a primary goal of any research process. The question is, does an organization need to engage an uninvolved third party in the process to accomplish this goal? It is true that someone not connected to the customer or the business is more likely to react objectively to the customer's comments and less likely to ask leading questions. However, the very factor that creates this objectivity — ignorance of the business and the customer — creates another problem: It is very difficult for an uninvolved third party to clarify the customer's comments to gain the most actionable data. One of the most common "next steps" after a third-party survey is to go back to customers and clarify what they meant.

The question boils down to, "Is it more effective to train the third-party interviewer on the organization's business, or is it more effective to train employees who know the business to conduct unbiased interviews?" Our choice is the latter. Customers are not as interested in an objective reaction to their comments as they are to energized action in response to their comments.

Beyond the inherent knowledge gap created by inserting a third party between a company and its customer is a far more insidious implication for long-term relationships. What message does a company send when it declares (as so many have) that a major business objective is "better customer relationships" and "increasing customer satisfaction," but the methodology used is sending a third party out to talk with customers? We submit that the message is "the relationship is in trouble."

The situation is analogous to an interpersonal relationship. It is only when communication between individuals has broken down that a third party (a counselor or therapist) is called. The counselor's first priority is reestablishing the lines of communication. If lines of communication between a supplier and a customer exist, however tenuous, far greater benefit is derived from customer/supplier communication than from "objective" third-party intervention. Why hire a third party to stay close to your customers for you when you can do it far more effectively yourself?

Assumption #3: Controlled data collection.

Reducing the variability introduced in the data collection process is a key goal of any research. The most controlled environments are those that simulate the laboratory. In survey research, the situations most analogous to the laboratory are those that expose all respondents to the same stimuli, ask or present the same questions in the same way and capture the response in the same manner.

Through the preparation of interview documents, the Quality Review process approaches this level of consistency. Selected individuals from an organization are trained to collect data using a structured interview guide and standard interviewing techniques. Using an interview guide and exposing all interviewers to the same training reduces variability of interviewing styles, ensures completeness of the information gathered, and reduces interviewer bias. In short, use of the interview guides increases reliability of the data.

"Getting as many managers out of the shop and talking to customers is the best research method a company can employ. It often seems to me that the conclusions reached after performing a lot of market research could have been reached by using common sense."

R. McKenna
Relationship Marketing, 1991

Within the Quality Review process, the format of the interview guides is standard and based on proven survey techniques. The benefit is that the content is specifically created by the organization that is ultimately responsible for using the information.

Challenging such basic assumptions about research methodology can be (and very likely will initially be) met with resistance from those with a vested interest in traditional marketing and customer research approaches. The reaction is understandable. Old allegiances die hard and new approaches engender anxiety. The Quality Review process is challenging not just a measurement system, but the way research people view the world and are viewed by the world. It moves the skill set valued by the company from the ability to crunch numbers to the ability to interact with customers. It may impact the status of individuals, determine who presents at major meetings and who goes to lunch with whom. Implementing the Quality Review process is not just an administrative decision — it is a decision with personal and personnel implications. Objections are inevitable. Resistance is predictable.

Let's Not Forget Hazel

We first met Hazel back in Chapter 6, where she was the focal point of a manager's objection to the Quality Review process. "Surely," he said, "we aren't going to *voluntarily* put Hazel in front of customers?"

Fear of putting the Hazels of the world in front of customers is representative of the kind of active objections that confront executives implementing the Quality Review process. Along with the "Hazel objection," four additional objections to the Quality Review process were raised:

- Our employees don't have time to talk to customers. They're too busy doing their jobs.
- Customers are too busy to talk to our employees.
- We've just introduced a new quality initiative and we don't want to introduce a new philosophy/language/training. We don't want to shift gears now.
- We're already doing a customer satisfaction survey. Why do we need this?

As noted in Chapter 6, our experience finds that these objections are symptomatic of a company's internal priorities, not external realities. Nonetheless, they are representative of the cultural change engendered by the Quality Review process. These five objections are addressed on the following pages.

We can't put Hazel in front of customers.

In every organization there are employees of whom it can be tactfully said, "They lack interpersonal skills." Nonetheless, they have no less a requirement for customer information in order to perform their jobs in a manner that meets customer requirements than do more polished performers. Training within the Quality Review process (Figure 6.2, Step 3, page 77) addresses the skills and behavior required to effectively interview customers. Implementation of the Quality Review process does not preclude an organization from using existing or supplemental interpersonal skills training. The point is that within the context of the Quality Review process, this training has a specific, job-related objective.

Our experience has taught us that employee attitudes toward customers are often the result of internally driven objectives. Recall the bus drivers from Chapter 2 who didn't stop for passengers because they had to keep to their schedules. Such internally focused objectives literally force employees to regard customers and customer requirements as annoying interruptions. The Quality Review process is a clear signal to the organization that the customer is number one. The combination of a clear focus on the customer, integration of the customer into business practices via data collected through the Quality Review process, and skills training provided as part of the review process implementation transforms Hazels into effective ambassadors for the organization and into vital links between your organization and its customers.

Our employees don't have time to talk to customers.

This is probably the most common objection to implementing the Quality Review process. Even after management of an organization recognizes and accepts that the mission of the organization must be focused on external customers, the objection arises. This behavior is not unlike that of the people participating in the play-ing-card experiment — even after sensing that a new model was operating, they clung to the old paradigm.

Of course, employees don't have *unallocated* time to talk to customers. If there's an employee anywhere in today's lean and mean organizations who has too little to do, hopefully he or she is updating a résumé. Finding work for excess resources is not a major management problem. Finding time to get everything done that needs to be done is the problem. Ah, but therein lies the rub.

What *needs* to be done? How many resources is your organization spending on activities that can be directly tied to the requirements and expectations of your customers? "There is nothing so useless," says management guru Peter Drucker, "as doing with great effi-ciency that which should not be done at all."[5] Merely accepting the logic of relationship management does not shift the paradigm or move your quality initiative into the Activist Component. A manage-ment system that drives behavior change throughout the organization must be implemented.

Freeing employees from other tasks to participate in the Quality Review process sends the message that the organization is truly focusing its efforts on meeting customer requirements. Adding the Quality Review process on top of other responsibilities sends a message that the old paradigm is still the status quo, and that the Quality Review process is just another management whim and it, too, shall pass. Commitment to the Quality Review process must be a commitment that is integrated into all aspects of the business (Chapter 10).

Customers are too busy to talk to our employees.

Research data discussed in Chapter 2 revealed that companies consider the relationship they have with their primary supplier contact as the most important factor in doing business. Given that finding, it is not surprising that customers who are too busy to be interviewed by a company's employees are few and far between. Companies implementing the Quality Review process are astonished that 98 percent of their customers are receptive to employee-conducted interviews.

"It is almost 40 years since I first advised executives to 'walk around' — that is, to get out of their offices, visit and talk to their associates in the company. This was the right advice then; now it is the wrong thing to do, and a waste of the executive's scarcest resource, his time. For now we know how to build upward information into the organization. To depend on walking around actually may lull executives into a false sense of security; it may make them believe that they have information when all they have is what their subordi-nates wanted them to hear.

"The right advice to executives now is to walk outside."

P. Drucker
For the Future: The 1990s and Beyond, 1992

But what about the other 2 percent? What if a customer insists he or she is too busy to participate? A negative reaction to a request for an interview provides an organization with a very important piece of data that could not have been obtained from a third-party survey (where typically, a significant percent of the sample does not respond). The organization has identified a *specific* customer who in essence does not recognize enough value in talking to his/ her supplier to make time for an interview. The customer is telling the supplier that beyond filling orders, the organization doesn't provide enough added value to make it worthwhile to take the time to talk to the supplier's people. Bad news? Yes. Valuable news? You bet. Before the customer is lost, the supplier has the opportunity to address the basic relationship issue, which is beyond the scope of simply collecting customer information.

We've just introduced a new quality initiative and don't want to introduce a conflicting philosophy/language/training.

As put forth in Chapter 1, the Quality Review process is not intended as a replacement for all of the tools, techniques and comprehensive quality management systems available today. There is no *one* quality solution for all companies.

The objective of the Quality Review process is to identify the requirements and expectations of the customers that those tools, techniques and quality systems must satisfy, as well as to provide a management system for ensuring that your organization's tools, techniques and quality systems continue to anticipate customer needs over time. As such, the Quality Review process is compatible with existing company initiatives. Implementation of the Quality Review process as an enhancement to information-gathering processes at companies that already have award-winning quality systems is evidence to support this.

We're already doing a customer satisfaction survey.

The Quality Review process does not preclude the use of traditional customer satisfaction measurements. Some companies choose to continue to conduct macro-level market research or surveys every few years. That is not bad, provided a company recognizes the

Figure 11.1 • **Differentiation of the Quality Review Process from Traditional Research**

CRITERIA	TRADITIONAL RESEARCH	QUALITY REVIEW PROCESS
Contract	Third party hired to implement and interpret measurements.	Organizations implement and interpret measurements themselves.
Focus	Inside-Out: prioritize improvement areas by analyzing internal work processes.	Outside-In: prioritize improvement areas by asking for customer priorities, and evaluating supplier performance.
Cycle	Measure on a project basis – every one or two years.	Measure on an ongoing basis – make it part of normal business activity.
Data Collection	Independent researchers collect customer data in confidential manner.	Employees collect internal and external customer data in direct, open exchange.
Customer Receptivity	Moderate – little confidence that information will reach appropriate target.	High – absolute confidence that information is reaching the right audience.
Analysis	Summarized across customers only.	Presented by individual customer and summarized across customers.
Response Rates	Variable, depending on method used.	Customers always responsive to employees of supplier.
Reliability of Data	Potential to misinterpret, act on unreliable data.	Opportunity to clarify, which increases reliability of data.
Feedback of Results	Management is the first to know – they control the use of the data. Organization receives data after significant time lag.	Employees who conduct interviews are first to know and can act on results immediately. Management is presented the global results and can focus action planning at a strategic level.
Communication	One-way communication tool.	Two-way communication tool.
Motivating Change	Data is top down, change is mandated – demotivates employees.	Ego investment in data collection process – motivates employees to improve.
Primary Objective	Measure customer satisfaction.	Measure and continuously improve customer commitment and supplier performance.
Business Results	Data collected for research purposes. Uncertain impact on business results.	Data collected for continuous improvement purposes. Direct measurable impact on business results.

limitations of the macro-view approach. A macro view of customer satisfaction provides *trending* information; it does not provide the specific, action-yielding, customer-by-customer information provided by the Quality Review process. An analogy may clarify the distinction.

Traditional customer satisfaction surveys are to customer satisfaction as stepping on the bathroom scale every morning is to losing weight. Over time, weighing yourself gives you some idea if you are losing or gaining weight, but it does not tell you *how* to lose weight. If you're eating the wrong foods or not getting enough exercise, you can weigh yourself all you want, but you're not going to lose weight. The Quality Review process is to customer satisfaction as researching proper diet and exercise levels are to losing weight. Armed with calorie counts, nutrition values, types of exercise most suited to your abilities, and the like, you can change your behavior to achieve your target weight. Maintain the behavior, and you'll maintain the weight.

An assumption in that last paragraph leads us to a final point about paradigms and the Quality Review process: A true paradigm shift not only changes *assumptions*, it also changes *behavior*. The latter is not optional. The Quality Review process not only challenges the assumptions of traditional research — the "why" things are done — it also significantly changes an organization's methods and measurement of customers — the "how" things are done. Figure 11.1 summarizes the differences between the Quality Review process and the traditional approach to customer satisfaction measurement.

In Chapter 10 we discussed the integration of the Quality Review process into the ongoing business practices of your organization. Integration, the ultimate goal of the quality movement, is the final step in implementing a quality initiative. To reach this goal, you must make the decision to change the behavior of your organization — to call a spade a spade, to shed the old paradigm and embrace the new.

Impartiality is what we ask as you consider the Quality Review process. We ask that you put aside the old concept of customer measurement and weigh the Quality Review process on its own merits. We ask that you consider your objections to the Quality Review process in light of the new paradigm of relationship management and external focus on customer commitment. It is that step — from old paradigm to new — that the growing number of organizations implementing the Quality Review process have made before realizing the full benefit of the process. Shifting the paradigm to an external focus has torn off the blinders and given these organizations an expanded view of their customers. It has allowed them to address issues their competitors are only vaguely aware of. It has positioned them to outdistance the competition. It has positioned them to win.

[1]   Thomas S. Kuhn. *The Structure of Scientific Revolutions.* Copyright 1962, 1970 University of Chicago Press, Chicago, IL.

[2]   Regis McKenna. *Relationship Marketing.* Copyright 1991 Regis McKenna. Addison-Wesley, Redding, MA.

[3]   McKenna. *Ibid.*

[4]   Theodore Levitt. "After the Sale Is Over… ," *Harvard Business Review,* September/October, 1983.

[5]   Peter Drucker. Quoted in *On Q, the Official Newsletter of the American Society for Quality Control.* October, 1991.

Acknowledgments

SHORTLY AFTER WE FOUNDED QUALITY INSTITUTE, ONE MEMBER OF OUR company's board of directors suggested that we write a book outlining our philosophy and the methodology we use in working to help our customers improve the quality of their products and services. His suggestion fueled a discussion that lasted for the rest of the board meeting.

We left that meeting energized about the idea of writing this book, yet perplexed as to how we would ever find the time and resources to complete the book and at the same time make sure our company grew and prospered during its early years. In the months following our decision to move ahead, we came to realize how fortunate we were to have friends and advisors to turn to when we needed assistance. We are very grateful to these individuals for their advice, counsel and encouragement.

When we decided that we would author the book together, we knew we needed help. While each of us fancies ourself a competent writer, we knew we needed continuity in the style of writing, someone to mediate between three headstrong personalities, and perhaps most importantly, someone to help add interest and drama to the central message in each chapter. Craig Westover of NCR Corporation filled each of these roles in a superb manner. His ideas, writing style and patience in dealing with us throughout the past year proved to be invaluable as we worked toward meeting each deadline.

Writing a book, as one author said, is a lonely endeavor. During the late evenings and weekends we were often accompanied by Sarah Dirksen, our able communications manager, who painstakingly read and edited each line in the book not once, but many times. We owe the layout and typesetting to Larkin Mead, a talented designer and desktop publisher. The illustrations provided at the lead of each chapter were designed by Susan Bartel, who captured the essence of each chapter in her creative sketches. Many of the research findings and quotations referenced throughout the book were provided by Maynard A. Howe, Sr., and Dorothy M. Howe, whose diligent review of the literature on quality proved very helpful during the writing process. The final pre-press proofings of the copy by Mary Gaeddert and Quality Institute's Jennifer Rock were extremely thorough and much appreciated.

The feedback we received from the many individuals who expended the time and energy to read and comment on our semi-final draft proved invaluable. Our thanks (in no particular order) to Carol Chase Baum, Paul Daulerio, Dick Dickinson and Tansukh Dorawala of Texaco; Gary Morris of Ameritech; Paul Noakes of Motorola; T.M. "Don" Buck of Fluor Daniel; Gordon Geiger and Tom Mauro of Cargill; Don Currin of Arthur Andersen; Chuck Poirier of Packaging Corporation of America; Frank Miller of Darling International; Joe Incandela of Thomas Lee Companies; Gene Bier of Enhanced Telemanagement, Inc.; business consultant Arnie Nicklin; and Maynard A. Howe, Sr., retired.

A special acknowledgment goes to Quality Institute's Kathy Schoenbauer for her contribution and careful analysis of the ideas in each chapter.

Finally, we wish to acknowledge our parents: John and Mary Gaeddert, and Maynard Sr. and Dorothy Howe, who taught us by example that commitment to a cause greater than one's self gives meaning and purpose to the concept of service.

Index

About the Authors

Roger J. Howe, Ph.D., Dee Gaeddert, Ph.D., and Maynard A. Howe, Ph.D. are the founders of Quality Institute International, Inc., headquartered in St. Paul, Minnesota. They have worked extensively with Global 1000 corporations, helping to improve the effectiveness of the customer and employee relationships they manage. Their expertise is in designing and implementing databased processes that are innovative, easy to integrate, and effective for achieving organizational objectives. Together and separately, they have written books and articles focusing on vital issues facing business managers today.